Sustainable Infrastructure:
Sustainable Buildings

Other titles in the Delivering Sustainable Infrastructure series:

Sustainable Infrastructure: Principles into Practice
Sustainable Buildings
Sustainable Water (coming soon)
Sustainable Transport (coming soon)

Institution of Civil Engineers

publishing

Sustainable Infrastructure: Sustainable Buildings

Elisabeth Green, Tristram Hope and Alan Yates

Published by ICE Publishing, One Great George Street, Westminster, London SW1P 3AA

Full details of ICE Publishing sales representatives and distributors can be found at: www.icevirtuallibrary.com/info/printbooksales

Other titles by ICE Publishing:

Sustainable Infrastructure: Principles into Practice (Delivering Sustainable Infrastructure series).
C. Ainger and R. Fenner. ISBN 978-0-7277-5754-8
Achieving Sustainability in Construction.
R. Dhir. ISBN 978-0-7277-3404-4
Intelligent Buildings, Second Edition.
D. Clements-Croome. ISBN 978-0-7277-5734-0

www.icevirtuallibrary.com

A catalogue record for this book is available from the British Library.

ISBN 978-0-7277-5806-4

© Thomas Telford Limited 2015

ICE Publishing is a division of Thomas Telford Ltd, a wholly-owned subsidiary of the Institution of Civil Engineers (ICE).

Commissioning Editor: Rachel Gerlis
Development Editor: Maria Inês Pinheiro
Production Editor: Rebecca Taylor
Market Development Executive: Elizabeth Hobson

Typeset by Academic + Technical, Bristol
Index created by Dr Laurence Errington
Printed and bound in Great Britain by TJ International Ltd, Padstow

Contents

Foreword

This timely book should be an essential reference for all parties concerned with maintaining the quality of our buildings with ever-decreasing impact on the environment. It explores the stewardship of our planet and possible interventions to the built environment for an effective response towards 'one planet living' of low or zero carbon, minimum waste, lower water consumption and a respect for the natural environment for a growing worldwide population of increasing affluence.

'Business as usual' in the design and procurement of the built environment will not enable societies, whether developed or developing, to meet the range of targets set by the European Union (EU) and others to reduce CO_2 emissions to 80% of those in 1990 by 2020, or in the UK Climate Act (2008) to limit CO_2 emissions to 20% by 2050.

In the sequencing of its chapters, this book proposes a successful route map for structural and civil engineers and other practitioners to close the gap between current understanding and the actual performance of the built outcome. It tackles the dilemma that buildings are complex and their occupiers more so. As a result, building performance is still largely unpredictable with regard to enduring sustainability. This book lays out the guiding principles of holistic and inclusive practice, followed by a useful explanation of the application of building physics and the impact of occupier behaviour. It explores in some depth why even buildings completed using the latest technologies and assessed against established certification schemes such as Leadership in Energy and Environmental Design (LEED) and Building Research Establishment Environmental Assessment Method (BREEAM) can fail to match their anticipated performance in practice as delightful, comfortable and efficient users of resource and energy. It is argued that to achieve actual targets of performance determined using benchmarking and post-occupancy evaluation studies, building control systems really do need to be ergonomically easy and intuitive to operate.

Critically, managing the process of inclusive procurement is very clearly explained in Chapter 5 in the context of flexible contracts between all parties. To my mind, along with full utilisation of building information modelling, this is the key to future success, and is it discussed very well in the context of the RIBA 2011 'green overlay' with its early workshopping between all stakeholders followed by an ordered stage-by-stage assessment of sustainability targets.

Many chapters conclude with useful checklists of what engineers can and should do to lead the delivery of sustainable buildings. Only by early adoption of the techniques and methods outlined in this book will both developed and developing societies meet the challenge of the Stern Report to reduce the end cost of mitigation and adaption of our built environment.

The built environment has a greater impact on our environment than any other aspect of life, so I strongly recommend all of us who care about this issue to peruse this book and its many useful case studies and references carefully. To play back the illuminating quotation from the explorer Robert Swan: 'The greatest threat to our planet is the belief that someone else will save it.'

We engineers must continue to act intelligently together with other practitioners if we are to sustain the quality of the built and natural environment. This book will help us to meet this challenge.

Michael Dickson
CBE FREng FIStructE FICE Hon FRIBA

Founding Partner Buro Happold Consulting Engineers and Visiting Professor of Engineering Design, Department of Architecture and Civil Engineering, University of Bath, UK

Series introduction

The books in this series are engineering books about sustainability – not the other way around. They are for practicing engineers, and other infrastructure professionals, who know generally of the challenges of sustainability, but have not read specialist books on the subject. The books are accessible and focused, illustrated with case study examples. They will help the reader to understand sustainability; to identify and apply the critical changes needed in each infrastructure sector to provide more sustainable solutions; and to embed them into our engineering excellence tradition of delivering high-quality infrastructure, on time, within cost and with good service – as quickly and straightforwardly as possible.

Asking the right questions, at the right time

Many engineers start work in the later stages of project delivery: feasibility, project detailed design, construction or operation. They may find it difficult to see how, and when, the sometimes vague language of sustainability affects what they do. But engineers do have responsibility, and it matters, because every infrastructure project contributes to, or detracts from sustainability. As their career progresses, they may get involved in the earlier stages of projects: project scoping, or business strategy and objectives – where they can have a wider influence. Whatever stage is being worked in, there are opportunities to deliver more sustainable solutions; but the biggest opportunities are often in the earlier stages of the process, and at the very start of each stage. Do not underestimate the difference that can be made even in those later stages: the best opportunities to deliver more sustainable infrastructure projects come by asking radical questions right at the start of each stage, before the pressure of 'just get it done' production takes over.

Each book, whether it be on buildings, water, transport or other forms of infrastructure, refers to the key principles set out in the first book of the series: *Sustainable Infrastructure: Principles into Practice*. The sector specific books ask how these principles can be applied to particular types of infrastructure, highlights with examples where they are relevant, and asks "What can engineers do?" They provide a summary of the core sustainability challenges each infrastructure sector faces and focus on emerging good sustainable practice drawn from expert practitioners in each field. They give practical advice on "how to do it" and opportunities where sustainable solutions can be introduced.

The books only cover the sustainability aspects of infrastructure, so are to be used alongside standard engineering texts and methods. They are not aiming for complete coverage, nor to be 'balanced'; rather, they focus on things that most need changing, and will make the most difference, using latest best practice – to help engineers to deliver sustainable infrastructure.

Charles Ainger and Richard Fenner
Series editors
January 2015

About the series editors

Richard Fenner is a chartered civil engineer with over 35 years' experience of teaching awareness of environmental and sustainability issues to engineers, focusing on the maintenance and rehabilitation of water industry buried infrastructure assets, sustainable water management, and interactions between water, land and energy resource sectors. He is the recipient of several awards from the Institution of Civil Engineers, including the RA Carr Prize and the James Watt Medal.

Charles Ainger is a former Sustainable Development Director for MWH's UK Operations. He has extensive water and environmental engineering experience in 16 countries – from Europe to Asia. His particular interest is in facilitating effective innovation and change leadership in organisations moving to a more sustainable and carbon neutral approach. He is a recipient of an ICE President's Medal.

Richard and Charles are joint winners of the ICE George Stephenson Gold medal.

About this book

Background and purpose

We see a considerable gap in understanding and language between the concept of sustainability and the practice for actually achieving it in infrastructure. While many books have been published about sustainability in many aspects of the built environment, few of them have approached it from the time-limited, output-driven perspective of the practical infrastructure engineer/ practitioner. This book series – *Delivering Sustainable Infrastructure* – aims to fill the gap. It applies a common set of sustainability principles to practice in different infrastructure sectors. The series so far has covered water, buildings, transport and waste management, and more volumes may be added. Each volume takes a similar approach, and applies the common set of principles and practice that was developed in the first overarching book in this series, *Sustainable Infrastructure: Principles into Practice*.

This book

This is the sector book covering buildings. Much has been written about sustainable building design and procurement, and this volume does not seek to revisit the same ground. Rather, the focus is on providing a route map through this wealth of guidance, fitting it into the common set of guiding principles outlined throughout this series, and on providing practical and effective advice on how, when and what engineers and other professionals involved in infrastructure-related buildings need to consider.

Many of the systems covered by the term 'infrastructure', while often complex in their design, construction and operation, are designed to operate within clearly understood and predictable parameters. Buildings are arguably the most complex, unpredictable and interdependent of the sectors covered in this series. Firstly their performance is strongly influenced by the behaviour of their users. Secondly, they are made up of many inherently complex systems layered in a very restricted space, and this suggests the need for a high degree of integrated design, construction and operation which, unfortunately, is seldom the case. Finally the traditional procurement structures and professional responsibilities involved in many building projects result in a fragmented and short-term perspective, which is unhelpful and limiting when it comes to improving a project's economic, social and environmental performance.

This book combines practical experience with academic rigour to provide a summary of good practice and identify sources of information that will be helpful to those involved in the design and procurement of buildings.

The authors are all specialists in their respective fields, and the book contains examples and case studies from a range of projects worldwide to illustrate the technologies and approaches covered. For simplicity, and given the focus of the series, the term 'engineer' has been used as shorthand for all practitioners involved in the procurement of buildings, including architects, surveyors, constructors and others, many of whom may be equally, if not more, influential than engineers in the design and construction of a building. The most important thing is for all those involved to reach a common understanding of the objectives and principles involved so that they can work together as a team. This book provides a framework to allow this to happen.

This book should be used alongside a range of other sources of information on building performance, design and construction, and references are included for some of these but the reader may have his or her own preferred sources. It would be impossible to provide a complete technical coverage of the issues relating to sustainable performance and design. The authors aim to provide a practical and balanced view of current best practice, and a vision of the way forward in a sector that is rapidly changing both in its structures and processes as well as in the demands placed on its outputs and the solutions that are available.

Book structure

This book is divided into four parts:

Part I: Principles introduces the key sustainability issues and challenges relating to building design and construction. Chapter 1 provides the background and sets down a challenge to drive change in a sector that has a massive impact on our financial and physical health and wellbeing, as well as on the environment and the demands that our society and economy places on limited resources. Chapter 2 applies the core sustainability principles developed in *Principles into Practice* to the buildings sector, and thus identifies the priorities for change, focusing on the 'what' and the 'why' of sustainable building design and procurement.

Part II: Practice takes the reader through the best of current practice, outlining the 'how and wherefore' of addressing sustainability in all aspects of the building sector – context, construction, operation and measurement. Chapter 3 explores the issues of user behaviour and systems performance. It describes the key aspects of building performance that must be understood and followed through by all if the design aspirations are to be achieved. It illustrates why these are important and the consequences of a failure to address them or the linkages between them properly. Chapter 4 considers the ways in which a buildings performance is influenced by its occupiers and its wider context throughout its life. Timing is everything when it comes to maximising opportunities and avoiding barriers to sustainable design and performance, minimising costs and timely delivery of a quality product. Chapter 5, therefore, considers how to maximise the sustainability benefits through good management and timely decision-making. Finally, but equally important, is the need to ensure that decisions are properly informed through reliable measurement and credible assessment of likely performance and outcomes. Chapter 6 summarises the principles of measurement, and outlines the breadth of measurement and evaluation tools available to the practitioner. It describes their uses and limitations, and identifies the many benefits and pitfalls that these tools create.

Part III: Change guides the reader through the inevitable changes that will influence the sector in the future. These will increasingly impact on the way that buildings are commissioned, designed, constructed and operated. Traditional structures and practices will change, and those involved will need to adapt as a result. Chapter 7 focuses on the drive to a more collaborative and integrated approach to building design. It explores the key issues and the role of building information modelling (BIM) as a means of enhancing communication and data transfer between stakeholders. Chapter 8 explores some of the broader changes that will influence the way we measure success and quality into the future. Finally, Chapter 9 ('Envoi') provides a brief roundup of the challenges facing building professionals in the future. While the challenge may appear daunting to the reader, by working together with others involved in a project it is possible to make big advances with relatively minor changes, and by following the principles set out in this book you will be able to achieve more sustainable buildings with relative ease.

Part IV: Tools is in the form of an appendix, and lists some of the principal tools and methods available. Given the breadth of sustainability, and the wide impact and complexity of buildings, there is a confusing array of tools available, and it would not be possible or helpful to list them all. The key tools are grouped into broad categories covering the principal whole and specialist building performance needs. This list is not, and can never be, fully comprehensive, and is based on the experience of the authors. Many consultancies, manufacturers and other organisations have developed their own specialist tools to aid their work, and to help others to specify services or products. These are not listed, as they are either not freely available or are not independently verifiable, and it would not be possible to provide full global coverage. While this book is primarily aimed at the UK professional, it is recognised that many will be working in other countries worldwide, and the list includes many leading international tools. The changes outlined in Part 3 will also result in changes to these tools and the creation or abandonment of others. This list should, therefore, be seen as a helpful resource rather than a robust and up-to-date resource library.

This sector book makes many references to the overarching book in this series – *Principles into Practice*. While we hope that you will read that too, this volume aims to stand on its own; the general volume gives much more detail on the background to the underlying principles behind this series, and why they are important.

Different readers will find different chapters of more direct relevance to their work. Start by reading Part 1 but then, if time is limited, feel free to skim the remainder to give a general overview, before dipping into the detail in Parts 2–4 as your role, needs and interest dictate.

If you know what questions to ask and when, you will find it 'do-able' to play your part in innovating towards a more sustainable solution. Even if you do not see yourself as being in the driving seat of change, you have more capacity to influence outcomes than you might think. Small changes in approach can result in big improvements in sustainable performance and procurement practices. This book aims to help you identify the *what, how and when*. We hope that you will use this book and enjoy it, and feed your own innovative experience into the shared experience and knowledge that the buildings sector desperately needs.

About the authors

Elisabeth Green
Senior Sustainability Consultant

Elisabeth Green has served as an Executive Board
Member of the Institution of Structural Engineers, as
well as on the IStructE Educational and Sustainable
Construction panels. Elisabeth's recent project work with
Mott MacDonald has been in the Middle East, providing
sustainability consultancy on masterplans, buildings and
regulations. Elisabeth is the author of the sustainability
chapter in the *ICE Manual of Structural Design:
Buildings*.

Tristram Hope
Founder and Director of THiSolutions Ltd

Tristram has 28 years of hands-on experience as a
structural engineer and building designer, having worked
with design companies that include Building Design
Partnership, Andrew Russell Associates, Buro Happold,
and Arup. He founded technical consultancy
THiSolutions Ltd in 2009. Tristram is an elected Council
Member, Board Member and Trustee of the Institution
of Structural Engineers, for which he also chairs the
influential Structural Futures Committee. His particular
areas of expertise include interdisciplinary integrated
building design, and the coordination and facilitation of
international construction projects. He has recently
worked in association with Charles Ainger (series editor)
as the reviewing editor of the IStuctE publication
Building for a sustainable future: an engineer's guide.

Alan Yates
Technical Director of BRE Global's Sustainability Group

As Technical Director, Alan is responsible for overseeing
all of the work of BRE Global relating to sustainability
of buildings, communities, materials and components. In
particular, he has led the technical development of BRE's
BREEAM (BRE's Environmental Assessment Method)
methodologies for over 20 years, and has worked closely
with the Department of Communities and Local
Government to develop the Code for Sustainable Homes,
based on BRE's EcoHomes method. In the past he has
worked on a wide range of environmental- and
sustainability-focused research and dissemination
projects, including the development of BRE's
Environmental Management Toolkits for existing
buildings. Alan is an architect with over 25 years'

experience in the design and development of environmentally sensitive buildings. Much of this time has been spent working at a strategic level with the UK Government and construction industry and internationally to improve understanding, acceptability and performance relating to sustainability. He has been a member of the ICE Environment and Sustainability Board and the RIBA Sustainable Futures Group, and has contributed to many working groups on the future of regulation and standards in the area of sustainability in the built environment.

Additional contributions by:

Nick Baker
The Martin Centre, University of Cambridge

Dr. Nick Baker qualified in physics but has spent the majority of his professional life working in building science as a teacher, researcher and consultant.

He has recently retired from the University of Cambridge's Department of Architecture, where he was involved with several EU funded research projects. His particular interests lie in energy modelling, thermal comfort, natural ventilation and daylighting, on which topics he is widely published. Nick has written several books, including *The Handbook of Sustainable Refurbishment*, and contributed to others on comfort and sustainability.

His recent work has focused on the refurbishment of the existing building stock and on the impact of human behaviour on energy consumption in buildings.

Andy Ford CEng, DEng
London South Bank University

Professor Andy Ford is Director of the Centre for Efficient and Renewable Energy in Buildings (CEREB) at London South Bank University. His research interests concern the challenges and opportunities presented by the need to decarbonise our energy supply and create an enjoyable and liveable built environment. He has had a long interest in research, innovation and knowledge transfer, working as a research manager in the DTI 'Partners in Innovation' programme and serving on BRE's Modern Built Environment Knowledge Transfer Network (MBE-KTN) steering group since its foundation.

Andy was awarded the IMechE Built Environment Prize in 2008 and an honorary doctorate by Heriot Watt University in 2012. He is also a past president of the Chartered Institution of Building Services Engineers (CIBSE).

Richard Shennan

Group Practice Manager, Buildings, at Mott MacDonald

Richard Shennan is Group Practice Manager for Buildings at Mott MacDonald, and has global responsibility for implementation of a strategy that is based on the performance of the built environment and the wider outcomes in terms of social benefit. He is also Group BIM Champion, responsible for the implementation of the global BIM strategy, which is increasingly focused on delivering the value of through-life asset information management to asset owners. He was previously managing director of Fulcrum Consulting, which joined the Mott MacDonald Group in 2009 having demonstrated leadership in sustainable buildings for over 25 years. Richard is chair of the Association of Consulting Engineers BIM support and engagement group, and a member of the Chartered Institution of Building Services Engineers (CIBSE) BIM steering group. He is a chartered engineer, and member of CIBSE and the Institute of Energy.

Part I

Principles

Sustainable Infrastructure: Sustainable Buildings
ISBN 978-0-7277-5806-4

ICE Publishing: All rights reserved
http://dx.doi.org/10.1680/sisb.58064.003

Chapter 1
Introduction

Alan Yates

> We don't, in a sensible world, want to hand on an increasingly dysfunctional
> world to our grandchildren.
>
> HRH Prince Charles, 2013

This book is part of a series covering key aspects of the delivery of sustainable infra-structure. While the series is aimed primarily at civil and building engineers, it is also of importance to the other professionals and stakeholders involved in the planning, designing, developing, operating and maintenance of our built environment. The first volume in the series – *Sustainable Infrastructure: Principles into Practice* – sets out the case for addressing sustainability in infrastructure, the elements of sustainability to be considered, and the importance of this issue to engineers and others in the future development of our built environment. The subsequent volumes explore the drivers, opportunities and risks involved in a number of specific sectors. This volume deals with buildings, their immediate surroundings and their relationship to the broader urban context, and explores ways of designing them so that they contribute to a more sustainable built environment.

1.1. Buildings' role and impacts

The buildings that we construct, occupy and adapt are a major resource, and indeed are critical to our success as a society and economy. They also consume large quantities of resources and have a major impact on our health, wealth and the environment.

- Given the amount of time that the average person spends in and around buildings, it is important to ensure that the environment they create inside and outside meets our physiological and psychological needs now and into the future (Box 1.1).
- Global and national economies are highly dependent on property values for both their scale and resilience (Box 1.2).
- Buildings are responsible for a significant proportion of our total consumption of resources. Global demand for resources such as energy, metals and rare minerals is increasing rapidly with global economic development, and it is vital that we make the most of these resources to ensure that future demands can be met cost-effectively (Boxes 1.3 and 1.4).
- Whether intentionally or not, buildings often make powerful statements about our individual and corporate attitudes to social responsibility, ethics and the environment.

Box 1.1 Impact of buildings on health and wellbeing

Buildings have a major impact on the health and wellbeing of occupants. It is commonly quoted that, on average, we spend over 90% of our lives inside buildings and travelling between them.

In the USA, typical indoor pollution levels are five times those outside, and may be as much as 100 times as high in some buildings. Given that asthma is the third largest cause of hospital admissions of children aged under 15 years in the USA, and is increasing rapidly, this is a cause for significant concern (US EPA and US CPSC, 2014).

It is not surprising, therefore, that, over recent decades, interest in more sustainable buildings has grown rapidly. Many buildings are claimed to represent high levels of sustainability in their design and construction. Some succeed but others fail to live up to these expectations. There are many reasons for this, but by setting out a range of coordinated and synergistic principles, considerations and processes, this book will help clients, their designers and others involved in the design and procurement of buildings to ensure that the breadth of issues are properly addressed and the risks of failure are minimised.

Box 1.2 Contribution of buildings to the economy

Buildings make a major contribution to the value and the stability of the economy. Compared with other aspects of the economy, the economic contribution from buildings is very significant (Figure 1.1).

Figure 1.1 The relative value of UK property and the UK equities market in 2012 (British Property Federation, 2013)

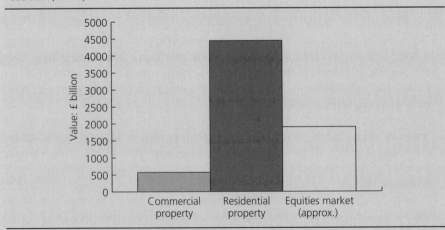

Box 1.3 Energy use in buildings

Globally, 40–60% of total national energy consumption results from heating, lighting, ventilating and servicing buildings, making them a major contributor to individual and business energy costs as well as to emissions of a range of pollutants including carbon dioxide, the largest single contributor to global warming and hence climate change.

Energy use varies between countries and regions because of variations in climate, generation mix and social expectations (Figure 1.2).

Figure 1.2 Split of energy consumption in the UK and the USA (BRE, various sources; US Department of Energy, 2014)

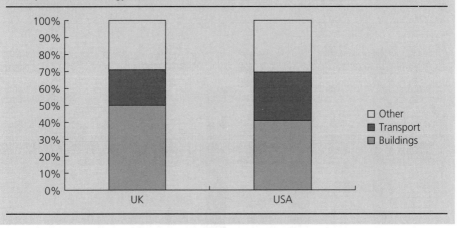

1.2. The sustainability challenge

Sustainability is usually described as a balance between economic, social and environmental issues. Resources are sometimes separated from environment, to represent the specific issues relating to them over those concerning the wider impacts of a building on its surroundings. This description of the concept of sustainability is frequently criticised, especially in terms of the equal balance between economic, social and environmental factors implied by the classic Venn diagram representation of three overlapping circles. What is widely accepted, however, is that these three strands are often overlapping, and each is dependent on the other two. A fuller discussion of these issues is given in the parent volume in this series (Ainger and Fenner, 2014, pp. 10–11).

The World Commission on Environment and Development (WCED), more commonly known as the Brundtland Commission, was formed by Javier Pérez de Cuéllar, former Secretary General of the United Nations, in December 1983. Its purpose was to unite countries in the pursuit of sustainable development. The Chair of the commission, Gro Harlem Brundtland, argued that:

the 'environment' is where we live; and 'development' is what we all do in attempting to improve our lot within that abode. The two are inseparable.

Box 1.4 Construction materials

Sales of primary aggregates in the UK were some 151 million tonnes in 2012, with a value of £1279 million based on estimated ex-quarry values (British Geological Survey, 2014).

The UK construction industry sent 12.55 million tonnes of construction, demolition and excavation waste to landfill in 2008, thus losing materials that could be reused in order to reduce demand for virgin materials. The UK Construction Strategy Board set a target to reduce this waste by 50% by 2012. Figure 1.3 shows the progress up to 2010 (the latest published data available at the time of writing).

Figure 1.3 Construction, demolition and excavation waste arisings in England 2008–2010. (Courtesy of WRAP/Strategic Forum)

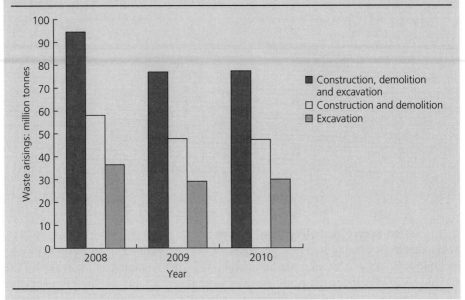

The following definitions relate to the first two key principles outlined in the introductory volume to this series (see also Table A1, page 248):

- *A1 – Environmental sustainability, within limits.* This principle establishes the environmental limits within which we need to operate to achieve a sustainable outcome.
- *A2 – Socio-economic sustainability, 'development'.* This principle defines the socio-economic goals that sustainability demands from us and the projects that we are responsible for.

Brundtland included 'social and political circumstances' in her definition of 'environment', and made it clear that 'development' places a clear responsibility on all countries to work to ameliorate 'our common situation'. The Brundtland Commission published its report *Our Common Future* in 1987 (United Nations, 1987) and coined the phrase 'sustainable development'. Although many have tried to refine that definition, it has endured due to its simplicity and accessibility. As such, it is still widely quoted today, and forms the basis of most discussions and approaches to sustainable development in use, including those described later in this volume:

> Sustainable development is the kind of development that meets the needs of the present without compromising the ability of future generations to meet their own needs.
>
> (United Nations, 1987)

Since the Industrial Revolution, the general approach to economic and social development and growth has been based on an ever-increasing application of technology and an unrestricted exploitation of available resources. This approach presupposes that the world's natural resources, from minerals and other materials through to energy and water, are effectively infinite. It also depends on the assumption that energy will continue to be available relatively cheaply. This approach has yielded significant rewards for developed countries, and others are increasingly seeking to follow in their footsteps using the past practices and historical precedents of the developed countries as a model for future success.

It would, however, be a mistake to think that this way of doing things can continue unabated, due to the following factors:

- Readily available fossil fuel resources are being rapidly depleted. While other reserves exist, and more will probably be found, they will be more expensive to exploit and are likely to require a greater degree of intervention to extract and process, with consequently increased impacts both socially and environmentally.
- Metals, rare minerals and other resources are being consumed at an unsustainable rate given known reserves, and global demand is growing rapidly. In a consumption-driven international market, demand will increasingly exceed supply, leading to significant increases in commodity costs. This phenomenon can already be seen to be occurring in the world's commodity markets.
- The Earth's climate is changing. At the same time, levels of carbon dioxide and other greenhouse gases that trap heat in our atmosphere are increasing rapidly. This strongly suggests a close causal link between atmospheric concentrations of greenhouse gases and average global temperatures. Regardless of the debate over whether or not these changes are man-made, they will change demands on the built environment over the lifetime of the buildings that are now being constructed and refurbished.
- Expectations of an improved standard of living, lifestyle patterns, business practices and technologies are continually evolving, impacting on the functionality that we demand from our buildings and infrastructure.

The energy crisis in the mid-1970s triggered upward pressures on energy costs, and these continue today as demand increases at both a national and international level, and energy resources become more expensive to exploit. This was an important trigger point for concerns about the energy efficiency of buildings, which resulted in changes to the Building Regulations in the UK in 1985. The subsequent tightening of regulatory standards led to the UK Government's target for 'zero carbon' buildings by 2019. The UK Government's previous 2016 emissions target for domestic properties was arguably made less stringent partially as a result of the policy of encouraging much needed house building by reducing performance requirements on developers which were being seen as burdensome. The changes also reflected the difficulties being experienced in driving rapid change in the house-building sector.

The UK Government launched the Energy Efficiency Best Practice programme operated by the Building Research Establishment (BRE) and the Energy Technology Support Unit (ETSU) in 1989. This initiative has evolved into programmes operated by a range of non-governmental bodies that exist today, such as the Energy Savings Trust for housing, the Carbon Trust for non-domestic buildings, the Zero Carbon Hub, and others.

More recently, public concern has increased over issues such as deforestation, water scarcity and ecological damage. This pressure led to the launch of BRE's BREEAM scheme for the assessment and certification of sustainable buildings in 1990, the Forest Stewardship Council's (FSC) timber certification scheme in 1993, and a wide range of other environmental labels and tools that are now available worldwide (these are covered in greater detail in Chapter 6 in this book).

Organisations, from constructors to property occupiers, funders and managers, have all come under increasing pressure to demonstrate formally their credentials as considerate and responsible organisations through corporate social responsibility (CSR) reporting and similar initiatives.

Buildings are highly complex systems, and finding a sustainable design solution can seem a daunting task, involving issues that fall outside the direct control of any individual stakeholder in the process. While this is undoubtedly true, engineers and others in the design team have a key part to play in identifying these solutions and working together to implement them. There are many competing demands to balance, and commercial and economic pressures will often make it difficult to tackle some areas. This book aims to make these decisions more straightforward by describing the underlying principles that need to be considered by the engineer and the broader design team, establishing responsibilities and opportunities for addressing the issues, identifying the links between different considerations, and describing the tools and techniques that are available and how best to make use of them.

1.3.   Sustainability in buildings – the way forward

Achieving sustainability in buildings is not primarily about absolute performance. Ultimately, sustainable design is more about adaptability, resilience and user understanding than it is about predicted design performance based on assumed use patterns and user behaviour. Concepts such as zero waste or predicted energy consumption are invaluable in ensuring that the possibility of good performance exists; however, they cannot guarantee

Figure 1.4 Designing for extreme environments – the Halley IV Antarctic research station. (Courtesy of the British Antarctic Survey)

that performance. A failure to understand this and to act accordingly has resulted in the apparent failure of many leading 'high-performance' designs.

Although complex and difficult projects in themselves, it is possible to design successfully for extreme conditions, as the parameters and needs are essentially predictable and quantifiable. For example, the new Halley IV Antarctic research station designed by Faber Maunsell and Hugh Broughton Architects for the British Antarctic Survey and the Natural Environment Research Council (NERC) represents a dramatic and highly innovative approach to designing a building for a very difficult environment (Figure 1.4).

However, it is nearer to home that difficulties are experienced when an attempt is made to impose a similar level of certainty on buildings where the environmental conditions and user behaviour patterns are more varied and so cannot be predicted with confidence. Greater flexibility and resilience is required to prevent unforeseen behaviours and other factors from frustrating design aspirations.

> You need to design systems to accommodate failure rather than eliminate it.
> By trying to be perfect, many visions of sustainability are quite brittle.
>   Jamais Cascio (2010), writer and ethical futurist

A set of underlying principles that will help in understanding and mitigating these risks is presented in Chapter 2 and explored in detail in Chapters 3–6.

History has shown that new resources and technologies can be developed to face new problems, and no more so than at times of political, economic and social stress. The economic, social and climatic changes that we are increasingly facing will make 'business-as-usual' solutions and practices less and less tenable. As a result, we need to adapt our approach to building design to cope with changes that are deeper, more rapid and possibly less predictable than any that have been experienced in the past. However, this does not mean a need to adopt a regressive or anti-advancement approach.

The technologies exist to enhance performance drastically, if only it were possible to start again from scratch with our built environment and cost was no barrier. Clearly this is not possible. However, major improvements in the performance of both our new build and existing building stock can be made once the commitment is made to invest sufficiently to ensure future savings in materials, energy and long-term asset values.

Historically, investment in property has been seen as both long term and dependable. In recent decades, investment decisions relating to property, especially within the commercial sector, have typically been taken on a much shorter-term outlook. This has several outcomes. It means that investments are required to pay back on relatively short timescales and long-term risks are not considered. The result of this is that upfront capital costs become more critical and often override sensible and practical life-cycle based solutions. It also means that buildings are designed for short-term needs and are less flexible and adaptable, being considered as expendable resources to be replaced rather than adapted. Finally, it can result in less concern for quality and durability, resulting in more resources being required to maintain them over their life.

A new building structure constructed to current standards will typically be able to last for a period of at least 60 years, and often much longer, but the design life specified by the client is often considerably shorter. Many components within a building, such as the services, windows and cladding systems, are likely to have a considerably shorter lifespan than this but decisions made in the basic design strategy will limit what happens when these elements are replaced during the lifespan of the building as a whole. Over a longer timescale, technological and functional requirements will change, and there is a need to ensure that the structure and built form is able to accommodate these as far as possible without major alterations being required.

There is, therefore, a need to consider more closely the whole-life impacts of a building, ideally across the full life cycle, including expected replacement of systems and components, and to consider future changes in use and available end-of-life options. This does not mean that all buildings should be designed for a long life, as this may not be functionally possible and some are required to meet much shorter-term needs. In all cases, however, steps can be taken in the initial design and construction to facilitate change, reuse and recycling wherever possible. This creates a need to identify a more realistic design life, identifying where continuity is likely. Designing for greater flexibility and adaptability is central to meeting differing functional requirements and climatic and social needs (Operational Principles 01.2 – Structure business and projects sustainably, 03.1 – Plan long-term).

The Climate Change Act 2008 requires the UK Government to reduce total national emissions of carbon dioxide to a level 80% below 1990 levels by 2050. On a shorter time horizon, EU legislation requires all member states to reduce carbon dioxide emissions by 20% by 2020. Buildings, new and old, will be required to play a significant part in achieving this. Occupiers will increasingly demand higher levels of performance as fiscal, commercial and other pressures force them to reduce impacts. This will result in major changes in expectation that impact on both domestic and non-domestic property markets. As a result, even buildings constructed to meet current regulatory performance levels will reduce in value more rapidly than in the past.

Running against this drive for change is the perceived additional cost of constructing sustainable buildings. While widely held, this view is increasingly unrepresentative of the actual costs of adopting pragmatic solutions as knowledge improves and technology costs decrease. It is true that super-low-impact buildings do cost more to build at current costs, although they may save far more in terms of operational and maintenance costs over their life cycle (Operational Principle O3.2 – Consider all life-cycle stages). However, many less radical solutions are cost neutral and, indeed, actually save cost, even in design and construction terms, as they avoid the use for expensive building systems and controls to maintain satisfactory internal conditions.

A common message from environmentalists and others is that we need to change our expectations of building performance. This may be true in terms of the institutional norms that govern so much of our property market; however, it is not reasonable to expect occupiers, investors and others to accept radical changes in occupant behaviour or quality to achieve lower impacts. To move forward a way needs to be found to meet current and future expectations more imaginatively and innovatively. The following chapters will help the engineer in understanding the issues and options, but in the end such innovation is down to the professionals and other stakeholders involved in a project and no book can provide a 'pattern book' of solutions to fit the full range of specific situations.

Technology undoubtedly has a major part to play in enhancing sustainability in buildings, as has always been the case. Construction may be less technology driven than some industries, but there are many opportunities to transfer knowledge and solutions from other sectors to help meet the challenges of the future and reduce the costs of higher performance. Dramatic improvements have been, and more can be, made in terms of building fabric through improved quality management, off-site construction techniques, innovative design solutions and the development of new materials, including the potential for nanotechnologies to change dramatically the properties of the materials that we use. Building services efficiencies have continued to improve in all areas, including LED lighting, heating and cooling systems, controls technologies and renewable energy generation. Advances will continue, and Part III of this book outlines some of these.

Technology can certainly help but more passive approaches also have a part to play. Internationally there is a vast array of traditional solutions to the problems of building environments that are conducive to human occupation in a wide range of climatic conditions. These will seldom be directly applicable to meet today's needs, but they do provide valuable models capable of being adapted to reduce the current high demands for energy and materials. Passive ventilation and cooling strategies are common in historic vernacular architecture, and have been successfully adapted to meet current needs and reduce demands in a cost-effective way.

In the end, higher-performance designs will need closer collaboration in design to ensure that all parts of a building and its services are supporting each other fully, and avoid any conflicts that all too frequently occur in current practice. This is covered in more detail in Chapter 7.

1.4.   Beyond the building

In this volume, the authors have taken the view that any study of sustainability in buildings must go beyond the basic building envelope, the systems and the activities that it accommodates. A building is usually part of a larger system, and opportunities and efficiencies are lost if it is considered in isolation.

This broader perspective is becoming increasingly important as we move towards national and international targets for reducing carbon emissions and energy use. An optimal solution for reducing carbon emissions is likely to include both energy demand management and efficiency within a building, and the broader efficiencies, including the use of low-carbon technologies, in the systems and infrastructure that supply or service it. The latter could be in the form of national/regional solutions as part of the supporting networks, or local community-scale systems for utility supply, transport or flood management.

This book will consider the role of urban master planning in achieving optimal outcomes, as these are important factors in reducing demand for energy consumption, internal and external environmental controls, social amenities and eco-support systems (Operational Principles O4.1 – Open up the problem space, O4.2 – Deal with uncertainty, O4.4 – Integrate working roles and disciplines). All aspects of building design have been considered but the focus has been placed on environmental and social impacts rather than the economic ones, as these are generally less well understood and are rarely considered in current design and procurement practices. This is not to say that the authors consider economic factors to be less important, and these are covered, especially in connection with life-cycle costs. The intention of this volume is to redress the balance so that all aspects of sustainable design can be fairly considered alongside each other, and a truly more sustainable outcome achieved as a result.

While the authors have focused primarily on the UK context, sustainability cannot be seen in isolation, so international factors and examples are used to illustrate common themes, as well as the differences, that need to be considered in other contexts. In terms of identifying and understanding the issues to be considered in designing sustainable buildings, this book is as valid internationally as it is for the UK.

1.5.   The way forward – developing good practice for sustainable buildings

While there are undoubtedly large challenges in designing, procuring and operating sustainable buildings, there is also a raft of new technologies, new approaches and even new mindsets that can be applied. Much has been written about sustainable and green buildings, and this provides a useful source for information and examples, but ultimately it is a change in mindset that is necessary to allow a new 'good practice' to be achieved. This practice will best be founded on a set of consistent sustainability principles, which are clearly relevant to the built environment. To provide this base, Box 1.5 takes the four

---

Box 1.5 The four Absolute Sustainability Principles applied to buildings (Ainger and Fenner, 2013)

Principle A1 – Environmental sustainability – within limits
- Minimise operational energy demand and maximise system efficiencies in meeting this demand.
- Minimise the carbon footprint of a building both at the procurement stage (embedded carbon) and in its operational phase (operational carbon).
- Choose the lowest practical life-cycle impact solutions for construction materials, building services and constructional details and built form.
- Aim to reduce waste through careful design and construction.
- Avoid unnecessary use of land resources, reusing land wherever possible and taking the opportunity to enhance land quality as a result of development.
- Minimise the use of water and other resources through careful design and improved manageability and understanding.
- Minimise the impact of development on ecosystems, valuing and enhancing the contribution that such systems make to environmental and socio-economic wellbeing.

Principle A2 – Socio-economic sustainability – 'development'
- Meet basic human needs for shelter, comfort, health and social interaction.
- Ensure inclusion for all, including the young, elderly, minorities, those with disabilities, and other vulnerable members of society.
- Seek opportunities to enhance community health and wellbeing.
- Minimise the risk of flooding through careful consideration of location and local surface water runoff attenuation systems to avoid impacts on others up or down stream.
- Create resilience to flooding, climatic events, functional and technological changes.
- Enhance skills and employment through design and construction.
- Source responsibly to avoid unnecessary environmental, social and economic detriment throughout the supply chain.

Principle A3 – Intergenerational stewardship
- Design, construct and manage along whole-life principles.
- Design for adaptability in function, technology and climate.
- Consider whole-life costs as a key part of the decision-making process.
- Ensure legibility in building design to facilitate ongoing operational performance, understanding and adaptability.

Principle A4 – Complex systems
- Consider sustainability holistically to avoid unforeseen consequences and limitations.
- Consider the building as a part of a broader system in terms of infrastructure and community.
- Identify the appropriate design life to minimise life-cycle impacts and costs.

---

absolute principles outlined in the first volume in this series and interprets them for the buildings sector.

These principles are explored in detail in later chapters, which include guidance, tools and examples. They illustrate that there is no reason why buildings should be built in the same way that they were a decade ago. Sustainable building design presents an opportunity for engineers, architects, suppliers and manufacturers to develop new skills, products and markets for their services, both in the UK and abroad. More than that, there is a responsibility for all professionals to take a lead in evolving the built environment to meet the demands of the future. As Robert Swan, the first person to walk to both poles, put it:

The greatest threat to our planet is the belief that someone else will save it.

Each chapter in this book addresses these key issues, and explores new ways in which familiar services can be delivered more sustainably. We hope this will provide a practical manual of things that can be done and adopted, rather than just a critique of the *status quo*.

There is a lot that engineers and other professionals can do to promote and encourage more sustainable design and procurement of buildings. The construction industry is diverse and includes many smaller organisations, and this can act as a major barrier to achieving this objective. Changes in procurement practices, including the growth of more integrated design and information-sharing systems such as building information modelling (BIM), will assist but it is necessary that we all play our part.

The opportunity exists to make dramatic improvements in the quality of the built environment, boost social–economic growth, raise ethical and responsible standards worldwide, and protect and enhance the local and global environment. It just needs all those involved to rise to the challenge and innovate. In the end, innovation and greater integration in design are the key to reducing costs, improving performance and contributing to the health and wellbeing of our society, our economy and everyone that depends on them.

We are not here to curse the darkness, but to light the candle that can guide us through that darkness to a safe and sane future.

John F. Kennedy

REFERENCES

Ainger CM and Fenner RA (2014) *Sustainable Infrastructure: Principles into Practice*. ICE Publishing, London, UK.

British Geological Survey (2014) Available at http://www.bgs.ac.uk/Planning4Minerals/Economics_1.htm (accessed 19/12/14).

British Property Federation (2013) *Property Data Report 2013*. Available at http://www.bpf.org.uk/en/files/reita_files/property_data/BPF_Property_Data_booklet_2013_spreads_web.pdf (accessed 30/10/14).

Cascio J (2010) article in *Green Futures*, January.

United Nations (1987) *Report of the World Commission on Environment and Development: Our Common Future*. Annex to General Assembly document A/42/427 – Development and International Co-operation: Environment. United Nations World Commission on Environment and Development, Geneva, Switzerland.

US Department of Energy (2014) *Buildings Energy Data Book*. Available at http://buildings databook.eren.doe.gov/ChapterIntro1.aspx (accessed 30/10/14).

US EPA (Environment Protection Agency) and US CPSC (Consumer Product Safety Commission) (2014) *The Inside Story: A Guide to Indoor Air Quality*. EPA 402-K-93-007. Available at http://epa.gov/iaq/pubs/insidestory.html (accessed 30/10/14).

Sustainable Infrastructure: Sustainable Buildings
ISBN 978-0-7277-5806-4

ICE Publishing: All rights reserved
http://dx.doi.org/10.1680/sisb.58064.017

Chapter 2
Guiding principles

Tristram Hope and Alan Yates

All for one, one for all.

Alexandre Dumas – *The Three Musketeers*

2.1.   Introduction

The saying 'the hardest place to start is with a blank sheet of paper' describes a scenario familiar to many designers. Experienced practitioners will usually have developed their own set of technical knowledge and philosophical tenets to which they can refer when embarking upon the design process. Some of this knowledge and philosophy is fundamental and widely shared, some relates to general methods used through the design process, and some is particular to the individuals involved in defining the design.

In the construction sector, solutions are frequently based on what has been done 'successfully' in the past, although over what period success should be measured and how it should be evaluated is rarely considered. Minimal back-analysis is carried out afterwards in the design and construction sectors to ascertain whether the as-built solutions represented the best outcomes possible.

It is important not simply to follow established methods by rote but continually to refer back to the principles behind them, in order to ensure that they remain up to date, relevant to their circumstances and the best available. While this takes time and effort, it is the only way to ensure that the outcomes will be truly successful and lasting.

2.2.   Fundamental principles

There are many ways to define the underlying principles for sustainable building procurement and operation. This book sets out an approach that is simple, flexible and structured. In order for low-energy, environmentally friendly, sustainable buildings to be delivered successfully, and then for them to operate as intended, the designer will need to be aware of these principles and adopt them throughout his or her role.

The principles outlined in this chapter are of a fundamental nature. They relate to the common good. As such, they leave little room for negotiation. Their relevance is universal, and their adoption is virtually a prerequisite to sustainable design. The principles used to structure this book address the fundamentals for optimising the sustainability of a building project through careful and considered design, specification, construction, commissioning, handover, occupation, post-occupation support, maintenance and refurbishment. They define the final product: the building in operation. These principles are as follows:

- **Think whole life** – A building exists to meet the functional needs for protection and enclosure and ought not in normal circumstances be considered to be an end in itself. The way in which a building is procured is critical to its sustainability and, therefore, its ultimate success. In addition, a building usually far outlives the initial use envisaged for it, and if well designed will have a lifespan that far exceeds this. Even when a structure can no longer be used as a building, it is important to consider how its constituent resources can be efficiently and effectively reclaimed and reused.

- **Understand the needs and context** – A building is a part of a system. Very few buildings exist in isolation but are dependent on the infrastructure and urban environment within which they are sited. In this context, the word 'system' includes not just the way in which the building is serviced and operates to serve its users; it also includes the local environment, ecosystems, supply chains and infrastructure to which it is connected.

- **Manage the process well** – The building process is typically most open to opportunity – and, conversely, to risk – at the 'blank canvas' stage. Careful management and guidance of the design, construction and handover processes can overcome the negative aspects, and truly successful outcomes can be achieved without the stress and heartache that are often experienced.

- **Recognise the benefits of innovation** – It is easy to continue to follow the path of least resistance by repeating what has been done in the past. 'It worked before, so why should it not work now?' In reality, no two situations are the same; the context is changing rapidly, and the solutions from the past will not work now and into the future.

- **Work collaboratively** – Buildings are complex, and few are designed, constructed and operated by one person. They involve input from many, and most problems occur when these people fail to coordinate their work properly. A truly collaborative working pattern can be productive, efficient and rewarding.

Each of these underlying principles is outlined in this chapter and explored further throughout this book. When combined with a positive and enthusiastic outlook, the application of these principles can deliver major benefits in quality, sustainability and recognition for those involved in the process. All are well tried and tested, and the following chapters give examples of successes and tips and guidance to support the reader in their implementation. They also fit well with the underlying principles set out in the companion volume to this book, *Sustainable Infrastructure: Principles into Practice* (Ainger and Fenner, 2014).

2.3.   Whole-life planning

Planning and designing for the building in operation and throughout its whole life, including its end of life, is a key aspect of sustainability. The aim should be to ensure maximum flexibility and adaptability in the basic structures and systems to allow for change without trying to predict future requirements and use resources to cater for needs that are unpredictable and may never be required or used.

The diversity of the building procurement and property sectors often creates a major barrier to achieving this. A few key approaches can be adopted at the design stage that can have significant benefit in reducing whole-life impacts arising from a building.

2.3.1 Life-cycle assessment and whole-life costing

Life-cycle assessment (LCA) and whole-life costing (WLC) provide invaluable tools to the designer and their clients to ensure that life-cycle considerations are being considered appropriately. They allow clients and their consultants to understand the options available and their potential impacts, and so inform their decision-making to reduce costs and resource use throughout the building's life. Considering these issues from the outset is the only way to optimise the performance over the full life of the asset. Neither approach is used typically in the building sector at present, although the benefits of doing so are becoming increasingly clear.

2.3.1.1 Life-cycle assessment

LCA is the process of assessing the range of environmental impacts associated with all aspects of any particular material or product, from the extraction of raw materials, through material processing, manufacture, transportation and delivery, operational use, maintenance and repair, and reuse, recycling or disposal. Some LCA methods also consider the human health impact of materials selection as well.

The process involves the initial compilation of an inventory of material and energy inputs and outputs into the environment. These inputs and outputs are then evaluated to determine their potential impacts, and the outcomes can then be assessed as part of an option appraisal process to inform decisions as to whether to use or avoid particular materials or products. These data are key to the process, and it is important that they should be comparable if the approach is to successfully highlight the relative merits of different choices. Careful consideration is needed of the data credibility, sourcing and boundary conditions applied, to ensure that this is the case. This is a difficult and arduous task for the building designer to undertake individually.

The compilation of the material and energy inventory and the assessment of the inputs and outputs has become a specialist field, and a range of proprietary databases and LCA products and services are available to the designer. A selection of these is listed in Part IV of this book. They provide building designers with the ability to use LCA in their design processes in a clear and transparent way, without needing to get embroiled in the technical data collection and consistency issues that are critical to carrying out such a review successfully.

2.3.1.2 Whole-life costing

WLC is defined by the UK organisation Constructing Excellence on its website (Constructing Excellence, 2014) as 'the systematic consideration of all relevant costs and revenues associated with the acquisition and ownership of an asset', and is a means of comparing options and their associated cost and income streams over a period of time.

Initial capital and procurement expenditure (CAPEX) and operational expenditure over the life of the asset (OPEX) are taken into account. Initial costs include design, construction, materials and components, installation, purchase or leasing costs, fees and charges. Operational costs include rent, rates, cleaning, inspection, maintenance and repair, replacements and renewals, energy and utilities use, dismantling and disposal, security and management.

Operating costs should be considered over the period during which the current functional requirement for the building will be in place, as it is not possible to predict future uses in most cases. An exception would be the design of venues for the London Olympics 2012, where legacy uses were considered as a part of the initial procurement, given the particularly brief duration of their use for the Olympic and Paralympic Games.

Design life, durability, adaptability and maintenance are all important when considering the built-environment legacy that will be left for future generations, as successfully demonstrated during the organisation of the 2012 London Olympics (see Box 2.1).

Over all but the very shortest of timescales, operational costs generally significantly outweigh initial costs, and of these approximately 50% can be attributed to unforeseen maintenance and repair work.

'Cost' in this context has historically been taken to mean financial cost. However, following the progressive development of increasingly sophisticated modelling techniques, it is now possible to carry out whole-life assessments using carbon as the currency, as a means of evaluating the long-term environmental impact of planned developments. It is also possible, although usually rather imprecise, to include costs of an indirect nature, such as health and wellbeing, staff retention and recruitment, business operation and image-related costs, which can be very significant for the many stakeholders involved in a building project.

Box 2.1 Legacy thinking

The organisation of the 2012 London Olympics involved an unprecedented level of legacy planning to ensure that the benefits of the initial investment would be maximised over the years to come. Care was also taken to ensure that the structures provided to facilitate this one-off world-scale event could either be removed or appropriately adapted to serve subsequent functions. The basketball arena (Figure 2.1), which was dismantled and removed, and the Olympic stadium (Figure 2.2), which was designed for the reinforced concrete lower bowl to be retained and the structural steel upper tier and roof to be removed, are examples of this thinking.

Figure 2.1 The 2012 London Olympics basketball arena was designed to be dismantled after the games and the components reused. (Courtesy of Wilkinson Eyre Architects)

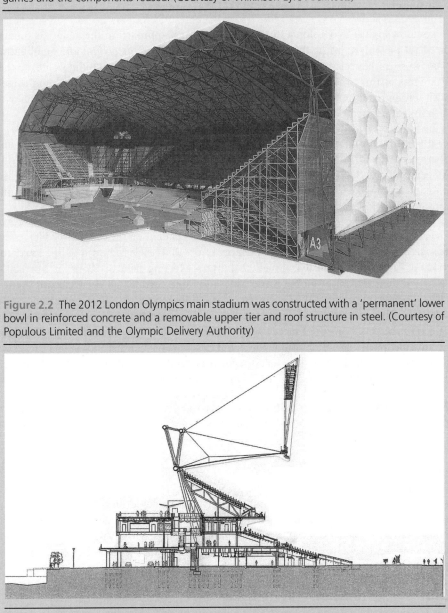

Figure 2.2 The 2012 London Olympics main stadium was constructed with a 'permanent' lower bowl in reinforced concrete and a removable upper tier and roof structure in steel. (Courtesy of Populous Limited and the Olympic Delivery Authority)

2.3.2 Design for deconstruction

Given the limited nature of Earth's natural resources and the need to consider the embedded carbon and embodied energy of a project over its full life cycle, it is important to consider how constituent resources can be reclaimed at the end of a building's life. A key component of this is the level of 'symbasis' of a manufactured product: how simply and efficiently the constituent primary materials can be segregated or reclaimed to allow their reuse or recycling.

It would be logical to develop a 'total asset inventory' approach to the materials present on site at the outset, listing those used in the construction of the project and highlighting those that can be reclaimed for reuse or recycling at the end of the building's life, so that awareness of this aspect can be maintained during the design and construction stages. Such an approach relies upon the ability to ascribe values (either in terms of money or carbon) to the materials present on site.

Useful information is provided in the University of Bath Inventory of Carbon and Energy, now incorporated in the BSRIA (Building Services Research and Information Association) guide to embodied carbon (Hammond and Jones, 2011), which lists whole-life average carbon dioxide (equivalent) figures for a range of over 200 commonly used construction materials.

The residual value of any material used is partly dependent on the ease with which it can be segregated for reuse or recycling; the more effort that needs to be put in, the lower the residual value. The G8 Environment Ministers Meeting held at Kobe in May 2008 discussed the problem of global waste generation and formulated the Kobe 3R Action Plan, agreeing on behalf of the G8 countries to action policies to reduce, reuse and recycle materials. The discussions recognised the problem of combining materials for use, and agreed to the development of more energy-efficient technologies to facilitate segregation. Meanwhile, particular attention at the detail design stage can help to enable the segregation of materials through designing for the disassembly of buildings at the end of their life – as illustrated in Figure 2.3 – rather than using the usual 'wrecking ball' approach.

2.4.   Understanding the needs and context

The success of a building is very strongly dependent on its context and the degree to which a design has taken this into consideration. Context is wide ranging but in terms of a building it includes: the locality; local and regional climatic, environmental and geological conditions; building physics and local vernacular; materials availability, supply chains, performance, resilience and the associated socio-economic impacts of materials procurement such as employment, skills, equity and resource stewardship; use patterns and occupant behaviours; and many more. Ideally, a sustainable building should sit as lightly on the ground as possible, with minimal impacts. However, this is usually not achievable, and significant impact is unavoidable. In such cases, impacts should be mitigated wherever possible.

2.4.1   Finite resources – living within limits

The Earth is finite in terms of the space and resources available, and an acceptance of the implications of this simple fact is a fundamental driver towards greater sustainability in our built environment, including the buildings that we construct and occupy.

This understanding has been around for many years; the seminal document *A Blueprint for Survival* was published in January 1972 (Goldsmith *et al.*, 1972), and highlighted the

Figure 2.3 Accessible bolted joints allow disassembly of structural frames. (Courtesy of Tristram Hope)

situation as it was then in stark terms, stating that: 'Governments are either refusing to face the relevant facts, or are briefing their scientists in such a way that their seriousness is played down'. It warned that 'we shall inevitably face the exhaustion of food supplies and most resources, and the collapse of society as we know it' unless the disruption of ecological processes was minimised, the human population stabilised and society in all its many parts accepted that the modern industrial way of life, stemming from an ethos of continuous expansion and growth, was fundamentally unsustainable.

While some might say that this doom and gloom perspective is overly apocalyptic and alarmist, the underlying warning remains based on sound science. We need to find ways that allow us to develop economically, physically and socially that will also enable us to avoid the outcomes envisaged by that document.

Our built environment has a greater impact on the environment than any other single aspect of life, and as such it is incumbent upon all construction project participants to be mindful of the resources needed to deliver, operate and maintain the buildings that are constructed, and to be judicious in their specification and management. Social and economic ambitions need to be serviced in a way that ensures the Earth's environmental limits are not overstretched.

Even on the small scale of individual projects, the individual impact of which may be minute but the cumulative effect of which is huge, building designers need to remain conscious of the overall situation and tend towards solutions that are flexible, adaptable and renewable, and which will have a minimal environmental impact.

2.4.1.1   One planet living – fair distribution
A helpful summary of straightforward, easily applied lifestyle changes is provided in the guide *One Planet Living* (Desai and King, 2006). Targeted primarily at a 'developed world' audience, it outlines a simple course of actions that, if followed, would have a significant carbon footprint benefit.

Not all the thinking described lies within the sphere of influence of the construction sector, although several significant elements do. Targeting zero carbon in delivery and operation of buildings; reducing both construction and operational waste to an absolute minimum; using sustainably sourced, reusable or recyclable materials; minimising water consumption and respecting the local natural environment – all fall directly within the day-to-day remit of building owners, designers, constructors and operators.

2.4.1.2   Continuous stewardship – the intergenerational perspective
A permanent long-term view needs to be taken if future generations are not to find them-selves in the situation of trying to achieve stable, sustainable conditions without adequate means at their disposal. In the context of the construction sector, this principle manifests itself in the adoption of policies such as life-cycle assessment, and the specification of materials that are certified to have come from responsibly managed sources.

The Forest Stewardship Council (FSC) was an early example, launched in 1996, of an authoritative body being established to monitor and accredit sources of construction products that use timber in their manufacture (Figure 2.4). The timber industry has subsequently developed other standards, including the Policy for the Endorsement of Forestry Certification (PEFC) scheme. BRE Global has developed a standard (BES 6001) for the generic certification of the responsible sourcing of construction materials

Figure 2.4 Forest Stewardship Council – 'The mark of responsible forestry' – and a forerunner in sustainable source traceability. (Courtesy of the FSC)

and products (BRE Global, 2014), and a number of other specialist schemes and sector-specific schemes have been, or are being, developed in the UK and European market based on either the BES 6001 model or the one set out in BS 8902:2009 (BSI, 2009).

2.4.2 Maximising benefits from the local built environment and vernacular solutions

Historical legacy can have a major influence on the choice of the optimally sustainable solution. In developed countries there is typically a substantial existing building stock, so the opportunity may present itself to retain and refurbish an existing property, albeit a property that falls significantly short of current energy-performance requirements in its current form.

While it is important to address the issue of achieving a cost-effective and practical reduction in operational energy wastage, choosing to refurbish rather than build from new can yield very significant embedded carbon savings in foundations and superstructure. Savings in embodied energy arise from the reduction in the extent of heavy construction work and mass of materials needed to achieve the desired outcome. There might also be the chance to preserve an element of local heritage interest, which may be beneficial from a societal and landscape point of view.

Particular expertise is needed in the assessment of the structural condition of existing buildings, as well as a willingness on the part of the project team to engage with the constraints imposed by the existing layout, and therefore to accept a degree of compromise. While working within the limits imposed by the existing structure and built form can create barriers and limitations, many of the most sustainable building projects overall are refurbishments or partial refurbishments.

The 2007–2011 refurbishment of the Deutsche Bank headquarters building in Frankfurt is a good example, with around 98% of the original structural frame being claimed to have been retained and reused to deliver one of the world's most highly rated 'green' high-rise buildings.

Neither need it happen only once: the Manchester Central Convention Complex opened for business in 2009, making it the third time that this magnificent structure has been retained and adapted for reuse. It was originally built as a main-line terminus station, which operated from 1880 to 1969, and was subsequently refurbished for use as the GMEX Exhibition and performance venue between 1982 and 1986.

In developing countries the situation is typically very different. The problem is frequently the lack of infrastructure and materials, so sustainable design decisions are likely to revolve around making the best use of what is available, in terms of both materials and labour (e.g. the use of masonry construction incorporating the 'Hatil' feature, as described in Box 2.2). A common pitfall to be aware of in such situations is the desire to adopt 'modern' solutions that ignore vernacular wisdom in the rush to embrace the perceived benefits of the 'developed world'. This approach is certainly not sustainable, and indeed can have disastrous consequences.

Box 2.2 Vernacular wisdom – the Hatil

The earthquake in Ismit, Turkey, in 1999 led to several studies of seismic activity in the North
Anatolian Fault zone. Local vernacular construction came to terms with the earthquake threat
long ago through the development of the Hatil – a horizontal, timber, ladder-like feature,
built full width through the masonry walls of local domestic buildings (Figure 2.5). Hatils
perform the multiple function of providing a degree of lateral tying, while forming controlled
horizontal slip planes through the brittle masonry and acting as resilient shock absorbers to
help dissipate seismic energy. Many residential buildings constructed in the area using
'modern' reinforced concrete frame and concrete blockwork infill technology failed
catastrophically during the seismic activity in 1999. Ironically, some of these featured purely
decorative horizontal Hatil look-alike details, formed in the external wall render.

Figure 2.5 A vernacular low-rise masonry building, featuring the horizontal timber Hatil detail.
(Courtesy of Tristram Hope)

2.4.3 Consideration of context

Most buildings are unique in their context. This is most apparent in terms of their
physical site, where even minor differences in location or orientation can see radical
changes in ground conditions or exposure effects, but it is also the case in the broader
contexts of regional climate, the state of cultural and social structures, and the economic
status of the countries in which they are built. It should, therefore, be expected that any
particular problem may have different yet equally appropriate answers, depending on the
context. In addition, local availability of materials, skills and finance (both capital and
operational) will have a significant effect on the sustainability of a building.

Difficulties can be expected in trying to apply standardised solutions, so it is essential that project teams should be prepared to take the time necessary to validate proposals within their specific contexts, by asking questions such as:

- Are the materials available and at a realistic cost?
- Is the local labour market capable of building to the required standards?
- What will the short-term and long-term impacts on the local area be?

Considering context in this way is an important initial step in avoiding the delivery of a building that is inappropriate, and that can therefore never form part of a sustainable solution.

2.4.4 Mitigation and adaptation

As stated earlier, projects should aim to 'touch the Earth lightly' through reducing negative impacts that are likely to have a long-term effect on the environment, and indeed eliminating them if possible.

Mitigation is the process of reducing to a minimum any negative impacts, such as energy wastage or the use of high embedded carbon materials. Adaptation complements mitigation, and involves coming to terms with the residual problems of climate change, largely through altering behavioural patterns.

Given the evidence of anthropogenic climate change that now exists, it is increasingly incumbent upon project teams to engage with policies of mitigation of environmental impact, and to consider subsequently how they might need to adapt to climatic conditions that are in a state of change.

2.5. Managing the process well

Buildings are highly complex products, and the typical procurement structures, responsibilities, legal relationships and risks reflect this. As a result, there are many conflicting interests and priorities present on the average building project. This frequently impedes the process of getting best value for the client and the performance of the final product. Clarity in project-management structures, relationships and gateways, the joint adoption of common priorities, targets and objectives, and the sharing of information, ideas and risks across the project team are all key to a successful outcome.

2.5.1 Setting common goals

Sustainability can be regarded as an underlying philosophy in which context projects are undertaken. It can have a root-and-branch effect on project-related decisions, so it is important that the project team, including the client, funding body, design team and other consultants, decide at the earliest possible stage what their commitment to sustainability is, and where the priorities lie. Objectives and targets can be established that can then guide the whole process of designing and procuring a building by giving a clarity and focus that can guide the measurement of success at each stage. Monitoring of these objectives is vital if the desired outcomes are to be achieved.

This can be achieved through holding an 'initial sustainability workshop' at the outset of the project. The agenda for the workshop should include discussion items such as:

- understanding the terms 'sustainable design' and 'triple bottom line'
- benchmarking, measurement, monitoring and reporting of sustainability
- project participants – roles and responsibilities
- effective project team communications
- typical construction-related sustainability issues
- project scope
- project-related sustainability aims and aspirations
- discussion and agreement of project-related sustainability opportunities
- recording of agreed outcomes
- the sustainability Action Plan
- measurement and monitoring of performance against these outcomes
- responsibilities and coordination within the project team
- a schedule of sustainability review meetings
- next steps.

The outcomes of the workshop will need to be formally recorded and signed off by all participants, and will provide the basis of a 'sustainability action plan' for the project. The purpose of this is to ensure that a hierarchy of sustainability-related considerations is established and understood at the outset, and is formally embedded within the project brief. The sustainability action plan will also need to be reviewed periodically in subsequent project team meetings, and there should be clear responsibilities and methodologies identified for this. Following this process can often have a dramatic impact on the time and cost involved in designing the building, and may indeed significantly reduce both capital and operational costs in the long run, even though it may mean an earlier involvement in the design process for some stakeholders.

2.5.2   Adopt a 'service', not 'product', business model

Buildings have traditionally been valued as providing both a 'service' – as a suitable environment to carry out various activities – and also a 'product' – as fixed real-estate assets in their own right. While these are often seen separately, they are in fact closely related, and this link is rapidly growing stronger as building performance requirements and the economic impacts of poor performance and high maintenance/upgrading costs increase.

Focusing initially on the aims that are to be achieved, rather than on the means by which they may be accomplished, keeps the range of possible solutions more open in the critical early stages. It may turn out that a reorganisation of activities is really what is needed, not the construction of a new building to accommodate existing activities. The adoption of a service, rather than a product, business model can therefore have a major influence on the sustainability of the overall project.

2.5.3   Adopt a holistic approach

As the building design team takes a more all-encompassing view of issues that affect building performance, it becomes apparent that actions taken in one area usually impact on other areas. Many of the failures in buildings occur when solutions are driven by single issues and concerns without due consideration of the impact of these decisions on other aspects of performance, cost and time.

In extreme cases this can create unforeseen difficulties, and even negate the intended benefits. For example, improving building-envelope airtightness to minimise energy loss due to infiltration may also lead to internal air-quality problems and structural/material degradation. The design of a building form around a specific building-services solution may make it more difficult to adapt the building to take account of functional or climatic changes without significant additional cost and disturbance. The need to think more broadly is not only restricted to the individual building in question; adjacent buildings can also have an impact, and may themselves be affected.

It is, therefore, crucial that a holistic outlook is taken to building design. Physical interactions with the local surroundings may become significant, and so necessitate specific study at the design stage. These influences include local environmental effects such as downdraughts and local increases in wind velocities due to wind funnelling adjacent to tall buildings (Box 2.3), and the 'heat island' effect, whereby the night-time re-radiation of energy absorbed by buildings or dark-surfaced areas during the previous day prevents ambient temperatures from returning to normal levels. This results in a gradual localised build-up of temperature over several days.

There is also the potential problem of rapid rainfall run off from impermeable surfaces such as building roofs and façades, roads and parking areas, which can result in local drainage systems becoming overwhelmed, leading to local flash flooding. Consideration should, where possible, be given to the incorporation of sustainable urban drainage systems (SUDSs).

The likelihood of such effects may not be readily apparent at an individual project level, so the building designer needs to be aware of the possibility of their occurrence and take a proactive, holistic approach when checking the impact of a new development.

2.5.4 Design gateways

Most UK building project timelines are described in relation to the RIBA (Royal Institute of British Architects) Plan of Work (RIBA, 2013). This has been in existence for many years, and was updated in 2013 to bring it into line with current design/construction practices and, most critically, to ensure compatibility with the rapidly developing building information modelling (BIM) agenda.

Sustainable building design requires that key decisions are made at the right time with proper consideration of key issues at a stage when they can be made without detrimental impact on other factors such as cost, time and practicality. Building evaluation methods such as BREEAM (see Chapter 6) have recognised the importance of this by requiring the appointment and engagement of specialists at key gateway stages in a project, and provide guidance on the optimum timing for key issues to be considered.

The work on developing BIM in the buildings and infrastructure sectors is considering the appropriate gateway stages for the sharing of information, and this will influence

Box 2.3 Local environmental effects – Bridgewater Place, Leeds, UK

Wind funnelling adjacent to the 32-storey, 110 m high Bridgewater Place office building in Leeds (Figure 2.6) led to a fatal accident in March 2011, when a delivery vehicle was blown over and landed on top of a passing pedestrian. The subsequent coroner's enquiry in December 2013 made the recommendation that adjacent roads should be closed off when wind speeds in the area reach 45 mph (20 m/s). Remedial measures have since been proposed to prevent severe local environmental effects from recurring.

Figure 2.6 Bridgewater Place, Leeds – increased wind velocities around high-rise buildings can lead to difficulties in the surrounding area. (Courtesy of Tristram Hope)

design relationships and data drop points, as well as information-sharing processes in the future (see Chapter 7).

2.6. Benefits of an innovative approach

The most successful solutions in engineering and design have often come from looking 'outside the box'. Most of what we accept and expect as the norm today originates from the innovative thinking of past great minds, ranging from Brunel, the 19th century engineer, to Le Corbusier and Mies Van de Rohe, two of the 20th century's most influential architects. Given the challenges that face us through technological and economic changes and in providing for the ongoing economic and social development within the bounds of finite resources, it is clear that our solutions of the past will not continue to meet the demands of the future. While this statement may be a very general one, there are specific approaches to which it is worth specifically drawing attention.

The designer should give continual thought to the possibilities presented by changing and adapting solutions to meet specific challenges. These may be through enhanced performance, or through more efficient and cheaper processes. Perhaps the greatest need for innovation is in the structuring of relationships and the sharing of information. Both of these are covered later in this chapter.

2.6.1   Lateral thinking
The phrase 'lateral thinking' was coined in 1967 by physicist Edward de Bono (De Bono, 1990). It refers to the solving of problems by avoiding the conventional method of using linear step-by-step logic, instead favouring an indirect and imaginative approach, starting by identifying the essential nature of a problem, then following lines of thought that may not be obviously apparent. This can be particularly useful when attempting to reconcile multiple, and sometimes contradictory, points of view – a situation frequently encountered in the building design process.

2.6.2   New skills and competencies
The development of 'soft skills' in areas such as communication, negotiation, organisation and lateral thinking is of assistance when it comes to identifying possible design solutions and reaching consensus agreements. In the case of international projects involving teams from different countries and cultures, a working knowledge of several different languages is also an advantage, and may indeed be essential.

2.6.3   Technical innovation
Technologies are moving rapidly in terms of both building systems and materials. This book does not describe these, not least because they change continuously and rapidly to meet new opportunities and challenges. However, it is worth pointing out the need to evaluate technology options and explore their relevance to the issues raised by individual projects. Rapid advances are being made all the time in systems efficiencies, control systems, thermal performance of materials, integrated and off-site manufacture and microgeneration technologies.

Refer to activity in other countries and sectors for potential ideas on innovative approaches. This is typically easier for manufacturers than for designers to do; however, many good design and construction ideas have been adapted from other sectors that have already developed solutions to analogous situations and problems.

2.6.4   Think longer term
More relevant here is the innovation of process, as this is more generic across projects. The construction industry, like most others, is driven ultimately by commercial interests involving a balance of cost and risk.

Commercial pressures, especially the desire to demonstrate a rapid financial return on investments, frequently result in project specifications being driven by short-term objectives. The 'quick in, quick out' attitude typically goes hand in hand with a disregard for the longer-term social or environmental consequences, which in the case of construction projects may extend over many decades.

Intuitively, this approach cannot be sustainable, given the life expectancy of the average building both as a service and as an asset. The building designer should, therefore, be cautious, and clarify the motives behind projects that appear to be dependent on delivery within very short timescales.

Occupiers, landlords and their agents are increasingly aware of a shift in asset value, and are now demanding longer-term security than previously. This changes the role and responsibilities of the designer and constructor in relation to their clients, society more broadly, and the environment.

Sufficient time needs to be allowed prior to commencement to carry out due diligence work thoroughly, carry out impact assessments, quantify risks in a meaningful and credible manner, arrange and carry out interactive surveys in the local area, and undertake investigation work. This will inevitably involve some initial time investment; however, it will also minimise cost and time overall by avoiding the unnecessary pricing of unquantified risk, and ultimately will pay off through the delivery of a more sustainable solution.

2.6.5   Personal reference frameworks

Each individual in the team can develop his or her own 'personal reference framework' with regard to a variety of important issues that have a bearing on design. These issues, and the questions they give rise to, include the following:

- Cooperation or conflict?
  - Are you the sort of person who is happy to fit in with others?
  - Do you enjoy working with, or maybe adapting, other people's thinking?
  - Do you prefer to make your own decisions?
  - Do you enjoy making bold, individual statements?
- Consume or conserve?
  - Are you happy to use materials in whatever quantity may be necessary to perform a particular task conveniently?
  - Do you enjoy the challenge of using materials in the most sparing way possible, even if this means going to considerable extra trouble?
- Natural or man-made?
  - Do you believe in using materials sourced as directly from nature as possible, even if their performance may be inferior?
  - Are you happy to use a smaller quantity of a highly processed material in the belief that it will be more effective, or last longer?
- Risk or certainty?
  - To what extent are you happy to accept the unknown, while making allowances in your design to cater for the unexpected?

- – Do you prefer to check everything in advance and leave as little doubt as possible, so that your design can be highly refined?
- ■ Functionality or flight of fancy?
  - – Does it matter to you whether the form your building will take depends on the function it will have to fulfil?
  - – Are you happy to let a visionary flash of imagination dictate the appearance, even if this means functional compromises?
- ■ Proven or innovative?
  - – Are you happy to follow established precedents, work with known technologies, and stay within the bounds of accepted guidance?
  - – Do you enjoy doing something completely new, exploring the unknown and challenging stereotypes?

There are not necessarily 'right' or 'wrong' answers to questions such as these, but when working with others as part of a complex team it is of benefit for all participants to be clear about their views, and to be able to promote and defend them effectively.

2.6.6 Challenging orthodoxy: questioning the brief

Adopting an orthodox approach in design is a 'comfortable' option. Following established thinking and using proven methods and materials in a well-documented manner will usually minimise immediate risk and maximise certainty of outcomes. On the other hand, it could be argued that readiness to follow orthodoxy has led to the state of affairs that now exists, whereby decisive action is now needed to avoid further irreversible depletion of natural resources and damage to the environment. Deliberately creating the forum to review an orthodox approach, and question whether other less-established solutions might be better suited, is good practice.

A thorough review of the drivers behind a project can also lead to unexpectedly beneficial changes in direction. If a project design team is being commissioned, the assumption is likely to be that some form of building is required, so to ask whether a building is the right answer in the circumstances, or whether some completely different solution may be more appropriate, can be uncomfortable. Nevertheless, the careful testing of assumptions and questioning of the brief is advisable, and is to be encouraged in the interests of reaching the most sustainable outcome.

2.7. Integrated design and collaborative working

One of the biggest challenges for the construction and property sectors is tackling the negative impacts of fragmentation. This is achieved partially through forming relationships and partially through the enhanced integration of processes and data. The latter is covered in some detail in Chapter 5. The need for improved communication becomes increasingly important as solutions become more complex and changes happen more rapidly.

2.7.1 Complex systems – interdependence of different parties

As increasingly stringent demands are placed on the built environment in terms of its technical performance, dealing with complex systems becomes inevitable. The development of the design to minimise energy use while optimising internal conditions within the building, coupled with the process of constructing the building as efficiently and cost-effectively as possible, involves progressive interaction between many different parties. Repeated discussion, review, adjustment and compromise are typically required, and all parties involved will need to allow sufficient time for this.

2.7.2   Collaboration and choice of contract

As increasingly high expectations of performance are placed on a new generation of low-energy, environmentally friendly buildings, so the interaction between the traditionally distinct disciplines of architecture, structural engineering and services design is becoming more important. It is, therefore, essential to foster an open-minded attitude within the whole project team, from the client through to the caretaker. Good intercommunication, and the means and a willingness to achieve it are vital, as is a shared awareness of common goals. It is impossible to get the team (and, therefore, the resulting building) to reach the highest possible levels of performance unless there is an 'all for one' attitude and a determination to meet and overcome challenges together.

This thinking was proposed in Sir Michael Latham's 1994 report *Constructing the Team*, and reiterated in Sir John Egan's 1998 report *Rethinking Construction* (Construction Task Force, 1998). It is prerequisite to the pursuit of sustainable design, where many subjective and often contrary views have to be reconciled, and there are no 'perfect answers'.

The choice of contract is an essential factor in facilitating sustainable design. Historically in the UK, construction contracts have been seen as a set of rules to be referred to reactively, when something has gone wrong. Such contracts include the Joint Contracts Tribunal (JCT) forms, which work with the inherent assumption of adversarial situations between two or more parties in disagreement. More recent collaborative contract documentation, such as the New Engineering Contract (NEC) suite of contracts, works with the assumption of shared interests, and therefore tends to encourage and facilitate proactive prior avoidance of conflict, thus enabling positive relationships between the various parties to be maintained, so saving time and money in the better interests of the common goal.

The recent provenance of the BIM approach to design (see Chapter 7) is also an important development in collaboration between parties. BIM is increasingly being used as an integrated enabler for the design, construction and operational management of buildings. To work well, BIM requires the consensual involvement of all project participants, both to contribute initially to the building model and then to maintain and make beneficial use of the information that has been gathered. It is, therefore, a process that presupposes collaboration, and that allows the possibility of all project participants to view 'bigger picture' information, beyond that for which they may consider themselves to be directly responsible.

Much work has been done recently by the Construction Industry Council (CIC) in the UK, in the development of the CIC Scope of Services documentation (CIC, 2007). This assists the project team by describing the many different activities that may be involved in the definition, construction and operation of buildings, such that responsibilities can be allocated in an open and equitable fashion at the earliest practicable stage.

2.7.3   Interdisciplinary integration

The multidisciplinary role of the architectural engineer has long been recognised in the USA. This is perhaps due in part to the wide range of climatic conditions encountered across the North American continent and the desire for a generic America-wide set of design solutions. This has led to an emphasis being placed on the development of closely controlled internal environments in buildings of all scales, which may not always be a valid requirement.

However, it has also meant that the teaching of a fully integrated approach to building design is more widespread in the USA than it is elsewhere, although it is now becoming increasingly necessary and appropriate universally. Some UK universities, such as the University of Leeds, the University of Bath and the University of Sheffield, have for many years been training architects and engineers together on integrated courses for part of the curriculum, and it may well be necessary to increase the adoption of this approach both for those joining the professions and for those already practicing within them. The professional institutions have a major part to play in achieving this, and hopefully they can avoid the vested interests that have traditionally got in the way of a more integrated and value-driven design team.

Many larger design consultancies have adopted a more multidisciplinary structure. While this has had mixed success in improving the holistic consideration of issues across the design process, it is a step in the right direction. It may also provide a useful model for closer integration between different organisations, provided that the systems are in place to allow enhanced integration and data sharing.

Coupled with this, the rapidly increasing demands and effects of humanity in relation to our planet's fixed physical limits are progressively becoming recognised as subjects for serious global concern. Correspondingly, there has been increased commercial pressure on the manufacturers of mechanical equipment for buildings to improve the environmental performance of their products. As mechanical and electrical plant efficiencies are steadily bettered, so the fuel-wastage margin will gradually be reduced, to the point where, in order to meet increasingly stringent environmental performance requirements, the effective design of buildings has to involve a consideration of aspects that hitherto have been considered insignificant.

Issues such as the thermal performance and permeability of the building envelope, and the ability of the building fabric to absorb and re-radiate energy, will start to become areas of importance to the overall design: low to zero energy performance becomes unfeasible to achieve without a consideration of these factors. The historic key criteria of aesthetics and spatial proportion find themselves jostling for a place alongside considerations of thermodynamics and material properties, not to mention economics, societal benefit and environmental impact. The modern building designer therefore needs to be something of an all-rounder by way of training and expertise.

What can engineers do?

The extent to which engineers engage with the thinking described above is a matter of personal choice. However, it is clear that the influence of mankind is having a significantly damaging effect on the environment in which we live. These effects will continue to get worse until serious action is taken on a global basis to stop, and then to reverse, current trends of growth and consumption.

With this in mind, engineers are encouraged to:

- Be mindful of natural and human resources and be judicious in their use.
- Tend towards flexible, adaptable, renewable solutions.
- Allow sufficient time in design for interdisciplinary discussion, review and adjustment.
- Set economic, social and environmental targets, and benchmark and measure against them.
- Organise initial sustainability workshops, agree sustainability action plans, follow up with sustainability reviews and assessments, and adjust accordingly.
- Encourage the adoption of service, rather than product business models.
- Mitigate project impacts, then adapt to residual circumstances.
- Consider together the social, environmental and economic contexts.
- Clarify uncertainty by allowing sufficient time for investigative work to be completed.
- Plan long term; adopt a life-cycle assessment approach; design for de-construction.
- Understand and take account of 'vernacular wisdom' in the use of materials and design features.
- Recommend the adoption of a collaborative form of contract by the project team.
- Develop personal reference frameworks and be prepared to promote and justify them.
- Be prepared to challenge orthodoxy and question the brief.

REFERENCES

Ainger CM and Fenner RA (2014) *Sustainable Infrastructure: Principles into Practice*. ICE Publishing, London, UK.

BRE Global (2014) BES 6001 Responsible Sourcing of Construction Products. BSI, London, UK.

BSI (British Standards Institution) (2009) BS 8902:2009 Responsible sourcing sector certification schemes for construction products. Specification. BSI, London, UK.

CIC (Construction Industry Council) (2007) *CIC Scope of Services*. RIBA Publishing, London, UK.

Constructing Excellence (2014) Whole Life Costing. Available at http://www.constructing excellence.org.uk/resources/themes/business/wholelifecosting.jsp (accessed 30 October 2014).

Construction Task Force (1998) *Rethinking Construction*. Department of Trade and Industry, London, UK.

De Bono E (1990) *Lateral Thinking*. Penguin Books London, UK.

Desai P and King P (2006) *One Planet Living*. Alastair Sawday Publishing, Bristol, UK.

G8 Environment Ministers Meeting (2008) *Kobe 3R Action Plan*. Kobe, Japan.

Goldsmith E, Prescott-Allen R, Allaby M, Davoll J and Lawrence S (1972) *A Blueprint for Survival*. Penguin Books, London, UK.

Hammond G and Jones C (2011) *Embodied Carbon – The Inventory of Carbon and Energy*. BSRIA, Bracknell, UK.

Latham M (1994) *Constructing the Team*. Department of the Environment, London, UK.

RIBA (2013) *RIBA Plan of Work 2013*. RIBA Publishing London, UK.

Part II

Practice

Sustainable Infrastructure: Sustainable Buildings
ISBN 978-0-7277-5806-4

ICE Publishing: All rights reserved
http://dx.doi.org/10.1680/sisb.58064.041

Chapter 3
Understanding the building physics and behavioural principles

Tristram Hope
Nick Baker
Andy Ford

> The greatest enemy of knowledge is not ignorance, it is the illusion of knowledge.
>
> Stephen Hawking

3.1. Introduction

The objective of this chapter is to identify and explain several key physical and behavioural aspects, about which building designers need to have a working knowledge if they are to design buildings that will be sustainable in their initial construction and subsequent use.

The chapter starts with an exploration of the basic human needs for shelter and security, and the primitive and vernacular origins of building. Technical sections then follow, providing an explanation of the basic physics behind several of the main environmental functions of buildings.

The intention is that this generic knowledge will enable designers to avoid the implementation of inappropriate or fundamentally unworkable solutions, and thereby produce successful buildings.

3.2. Shelter and security

One of the many attributes of the human species is its ability to manipulate the immediate surroundings to increase chances of survival – in other words to seek and provide shelter from adverse weather conditions and security from attack by predators. So effective are humans in doing this that they have succeeded in colonising the majority of the land surface of the globe, in spite of climate extremes. It is worth noting that until only recently this was achieved without resorting to the use of fossil fuels.

Humans are not the only animals to have learned how to create shelter to improve the chances of survival, but two other primitive behaviours set them apart from the rest of the animal world: the use of fire and the development of clothing. Both of these can be used to control the thermal balance of the body, which is a crucial requirement for

41

survival. However, initially, man made use of what nature offered, retreating into caves to enjoy the stabilised ground temperature, and emerging into the sunlight for warmth when needed or seeking the shade when not. The next step was to modify the natural surroundings, hence the process of building. In this way primitive architecture developed, with the functional criteria closest to survival being the driving force.

3.2.1   Primitive and vernacular origins

As knowledge of shelter building was spread by direct observation, and survival became more certain, allowing the builder to respond to other more subtle influences – in particular social, cultural, and even symbolic and aesthetic factors. Primitive architecture thus developed into vernacular architecture.

Vernacular architecture represents a fusion of climatic and social functions, influenced by the physical characteristics, properties and constraints of locally available materials. Perhaps it is a subliminal recognition that basic survival requirements have been met, and that accepted social behavioural patterns have been appropriately respected, that leads to the continuing popularity of vernacular architecture in many parts of the world today.

3.2.2   Lessons to learn from primitive and vernacular architecture

Clearly, the changing functional requirements of modern-day building occupants mean that it is not appropriate simply to revert to traditional solutions. A century ago, the majority of people's time was spent outdoors. In today's world, the converse is true; light, heat, ventilation or cooling may be needed at any time throughout the day.

However, two very important lessons can be learned from traditional methods of building. Firstly, it is the integrated, holistic consideration of location and climate, the building and its occupants that has led mainly to sustainable solutions. It therefore makes sense to adopt a similar holistic, integrated, multidisciplinary approach to building design today.

Secondly, although the building physics may not have been explicitly understood by vernacular builders, it was an ever-present, evolutionary factor that served to drive the development of more economical and acceptable solutions. This took place over time, and a gradual development of technology was acceptable due to the slow rate of societal change. More recently, changes in global economic and social conditions have been far more rapid, such that it has become necessary to leapfrog evolutionary processes by developing a more technical understanding of the principles of building physics that underlie vernacular architecture, in order to be able to respond swiftly to the changing demands that are being placed on the built environment.

3.3.   Behaviour, lifestyle and adaptation

Bearing in mind the long-term success of vernacular architecture, it is not surprising that it is frequently taken as an inspiration for sustainable building design. However, building technology has undergone a dramatic transformation since the late 1700s, driven largely by industrialisation, urbanisation and population growth. Against this background of rapid change, vernacular architecture appears to represent a sense of unchanging timelessness. As a result, in certain aspects it appears to be archetypically and demonstrably sustainable.

3.3.1 The hierarchy of needs

While the laws of physics have not changed – and even current-day changes in climate are small compared with the range of climates for which vernacular architecture has provided satisfactory solutions – the behavioural lifestyles and social expectations of those who live in industrially developed countries have altered dramatically. This has come about largely through the progressive change in developmental emphasis from industrial manufacturing, to commerce, to service industries, to leisure.

The progression of lifestyle needs was recognised by Abraham Maslow in 1943, and in 1954 he represented these needs graphically in his diagram of the hierarchy of needs (Figure 3.1).

Manufacturing processes have developed in complexity in parallel with this, requiring increasingly controlled internal environments, in turn requiring further development in the sophistication of building technology, all of which has increased dependency on energy and/or natural resources.

3.3.2 Occupant interaction

However, through carelessness or simply out of ignorance, it is quite possible for building occupants to negate the good intentions of the building designer, who may have assiduously incorporated a whole range of 'green' features, by failing to operate the building in an appropriate way.

Figure 3.1 Maslow's hierarchy of needs. The section at the base of the triangle lists the fundamental requirements for survival; the higher sections demonstrate progressively more sophisticated 'wish-list' items that require becoming increasingly dependent on circumstances to be realised. The needs for shelter and security are in the bottom two sections

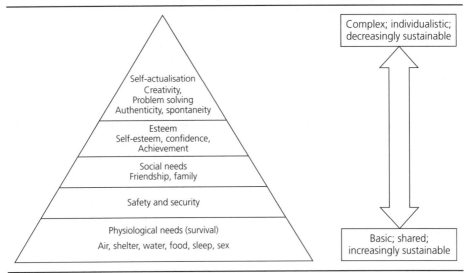

Behaviour can have a serious impact on building systems, and in some cases the failure of occupants to perform even simple intuitive actions such as opening windows to improve ventilation or turning down the heating to reduce room temperatures has led to carefully balanced, thoughtfully designed buildings being labelled as 'not working'.

It is therefore important that if we are to design low-impact buildings that use passive or mixed-mode systems to minimise their energy demand, we also need to ensure that those systems are understood by the occupants, possibly even to the extent of information sessions being provided periodically. Any controls that are provided, from technically-sophisticated, computer-based building management systems (BMSs) to simple manually operated levers, must be designed to be ergonomic and intuitively easy to operate. Many instances could be cited of sophisticated 'black-box' system controls that need a user manual to be operated correctly – only the user manual is not readily available, so the controls are not used.

3.3.3 Mitigation and adaptation

Al Gore's seminal documentary film *An Inconvenient Truth* (Paramount Classics, 2006) did much to help raise global recognition that the availability of natural resources is limited, and that global anthropogenic climate change is now a real and present danger. Adopting the combined strategies of minimising energy and natural resource consumption, while also adapting to changing circumstances, is essential. It therefore becomes necessary to reach acceptable lifestyle compromises.

As far as the built environment is concerned, much of this can readily be achieved through adaptive behaviour, whereby occupants interact with their environment in order to improve it, or, alternatively, change their own state to improve their perceived level of comfort.

3.4. Orientation, solar gain and shading
3.4.1 Façade configuration

Direct sunlight through normal glazing typically delivers between 0.25 and 0.75 kW/m^2, which if converted into heat can make a significant contribution to the heating load of a space, or in an overheating situation can add a large unwanted cooling load. The luminous intensity of direct sunlight is around 50 000 lux – 100 000 times more intense than the minimum illumination needed to read a newspaper.

Nevertheless, the position of the sun is specific and predictable in relation to geophysical references such as latitude, orientation, time of day and season. This means that the orientation, configuration and positioning of the main glazed areas in the façades of a building are within the ready control of the knowledgeable designer, as well as being of essential importance from the point of view of energy consumption and comfort.

Also, the term 'orientation' is often used very loosely when describing buildings. When considering orientation in relation to solar impact, it should always refer to the orientation of the principal glazed façade, and use the correct definition of orientation (i.e. the direction in which a line drawn at right angles to the façade will point).

3.4.2 Solar geometry

Solar geometry is a way of describing the position of the sun and hence the angle and direction of the sunlight. The building geometry should respond to this in order to maximise useful solar gain, or minimise unwanted solar gain. This requirement may change seasonally or even hourly, so to cope with this shading devices are employed. These may be fixed, responding only to the dynamic solar geometry, or moveable.

3.4.3 Sun-path diagrams and solar intensity

Software tools or sun-path diagrams can be used to understand the nature of the daily and seasonal movement of the sun through the sky, and its dependence on latitude. It is important to realise that the intensity of direct solar radiation is not constant but varies with solar altitude and atmospheric conditions. These variations are incorporated in the algorithms used in software and tables to predict solar intensities.

It is also important to remember that solar intensity is dependent on the angle of incidence of the radiation to the surface. The intensity falling on a surface at right angles to the beam is called the *normal intensity*. The intensity of radiation falling on a surface at any other angle is reduced by the cosine of the angle of the incident rays to the normal.

For example, in the northern hemisphere, if the sun has an azimuth of 180° (i.e. it is lying due south) and an altitude of 60°, the angle of incidence on a south-facing window will be 60°. This means that a normal intensity of 700 W/m^2 will be reduced by half (as the cosine of 60° is 0.5), to 350 W/m^2.

3.4.4 Solar gain and transmissivity of glass

Whenever solar radiation (both visible and invisible) is incident on a glazed aperture, some is reflected, some is absorbed (warming the glass) and some is transmitted. The greatest transmission of energy is in the visible region. The transmitted part (comprising visible and non-visible microwave, infrared and ultraviolet radiation) is then absorbed and reflected around the room. The absorption process warms the surfaces on which the radiation impinges, which in turn re-radiate energy at long wavelengths only, and lose heat to the room air by convection.

While short-wave visible radiation reflected by light-coloured surfaces is able to pass back out through glazed areas, long-wave radiation cannot. It is absorbed and heats up the room air.

It is important to realise that visible radiation, or 'light', transmits the same amount of potential heat as invisible radiation. This means that it is not possible get useful daylight without solar heat gain. However, it is possible to increase the efficacy of the daylight by using special glass that reflects or absorbs the non-visible part of the spectrum more than the visible radiation.

3.4.5 Shading

The functions of shading are to avoid the incidence of uncomfortably excessive levels of natural daylight or 'glare', and to control the ingress of solar radiation, without preventing the gain of useful levels of natural daylight, or compromising the view. This latter qualification is often overlooked, and the 'blinds down, lights on' scenario is frequently observed. As windows often play an important role in ventilation, the impact of fixed shading elements on airflow through the window opening can also be significant.

3.4.6   Fixed and moveable shading

There are two main categories of shading device: fixed and movable. Fixed overhangs, fins and louvres rely directly on their geometry to obstruct the part of the sky through which the sun passes, for all or part of the year. They cannot respond to day-to-day variations in weather, and the shading profile (how the transmission varies over the day and season) may not perfectly match the shading requirement. However, they are robust and can be integrated with the structure in the form of overhanging eaves or deep window reveals. They are often found in vernacular examples.

Moveable shading devices can be adjusted to match the shading requirement more closely, mainly by altering their geometry, but also by simply removing or deploying them, as in the case of a roller blind or curtains.

3.4.7   Positioning

It is important to note that where shading is to be used to combat solar gain it should be placed outside the insulated envelope of the building. If it is not, incident solar radiation will enter through the glazing and will be absorbed by the shading device. It will then re-radiate energy at longer wavelengths, which will be trapped internally within the building, thus defeating the objective.

3.5.   Heating, cooling and insulation
3.5.1   Conductive heat loss

Buildings lose heat by two primary mechanisms – the conduction of heat through the envelope, and convective loss due to ventilation. The former can be calculated by multiplying the area of the element being considered by its thermal conductance (commonly known as the 'U value') and the temperature difference between the air inside the element and the air outside (in degrees Celsius).

U values are published for a wide range of common building constructions, and range from about 6 W/m^2°C for single glazing down to 0.15 W/m^2°C for a roof insulated with a 300 mm layer of fibreglass. Combined layers of construction can be taken into account by adding the resistance of each layer. When performing this calculation, allowance is also made for the additional insulation afforded by the layers of still air adjacent to the inside and outside surfaces. When calculating the heat loss from a whole building, the losses through each element of the envelope are added together.

3.5.2   Ventilation heat loss

Ventilation (and unintentional air leakage, usually referred to as 'infiltration') can cause heat losses when warm air from inside the building is replaced with colder outside air. The ventilation heat loss is the heat required to warm the outside air up to room temperature.

3.5.3   Heat loss coefficient

The combined conductive and ventilation heat loss per degree Celsius temperature difference is often expressed as the heat loss coefficient. For a draughty and poorly-insulated three-bedroom house with single glazing and solid walls this could be as high as 500 W/°C, while for a modern low-energy house it can be as low as 80 W/°C. Note that, although the conductive losses can be reduced to very low values through the use of modern insulation materials, there is a lower limit to the ventilation heat loss, dictated by the requirement to provide a minimum amount of fresh air for the building occupants.

3.5.4   Degree-days, monthly and annual heat consumption

The heat loss coefficient can be used to calculate the heating energy required to keep the interior of the building at a set temperature, over a given period, using the unit of 'degree-days'. Degree-days are defined as the product of the time (in days) and the temperature difference between the outside and a set temperature inside the building (usually 18°C). Degree-day data are published for various parts of the UK and the world.

For example (multiplying by 24 and dividing by 1000 to obtain figures in kilowatt-hours), a three-bedroom house with a heat loss coefficient of 280 W/°C in the Thames Valley region of the UK (1896 degree-days/year) will require $1896 \times 280 \times 24/1000 = 12\,740$ kW h of heating.

3.5.5   Useful heat and boiler efficiency

The heat needed in the example above is 'useful' heat (i.e. the heat actually leaving the heat emitters within the building). In producing that heat from gas, oil or coal, some heat will be lost in the flue gases and maybe from pipework outside the heated zone.

The ratio of useful heat to delivered energy is referred to as the 'boiler efficiency'. Typical modern boiler efficiencies range from 75% to 95%, although older boilers may be much less efficient, especially when working at part load. The highest efficiencies are achieved through reclaiming heat by condensing the water vapour emitted in the flue gases.

3.5.6   Internal gains and solar gains

Not all the heat losses from a building have to be made up by auxiliary heating. Many activities, known as 'internal gains', such as cooking, lighting and the use of electronic machines, emit heat. Human occupants also produce heat. Provided there is proper thermostatic control in the room, these contributions will be deducted from the heat required from the boiler. It is important to note that, in almost every case, the use of equipment and lighting cannot be justified by its contribution to heating. This is because most heat-producing activities in buildings make use of electricity, which generates much higher carbon emissions per kilowatt-hour than heat generated from fuels such as natural gas.

Solar gains can also be useful in reducing the amount of auxiliary heating needed. With proper heating controls and the use of curtains at night, south-facing double glazing can provide a net gain of energy over a year. The usefulness of solar gains is crucially dependent on there being a demand for heat at the time when they are available.

3.5.7   Thermal mass

The usefulness of solar gains can be improved, and the problem of overheating diminished, by the use of thermal mass. Materials such as concrete, brick and stone are thermally conductive and massive, allowing large quantities of heat to be absorbed for a relatively small rise in temperature. Thus a sudden influx of solar gains to a room can, to some extent, be absorbed by the floor, the walls and the ceiling surface, thereby delaying the rise in room temperature. Stored heat will then be released back into the room as it cools down in the evening.

The calculation of the internal and useful solar gains is problematic. For this reason, computer-based simulation software has been developed to solve the highly complex heat-flow equations for short time steps, many times over.

To get the full benefit from structural thermal mass it is very important that gains have 'access' to the thermal mass. One further point to note about thermal mass is that the physical properties that improve the usefulness of solar gains and reduce overheating also render the building less suitable for intermittent occupancy, as the building is less able to respond thermally within short timeframes.

The sustainable performance of buildings usually requires several complementary strategies to be adopted in order to avoid the situation where benefits gained in one way are wasted in others. The German Passivhaus concept covers several important design aspects and is worthy of note (Box 3.1).

Box 3.1 The Passivhaus concept

The Passivhaus concept focuses primarily on reducing to a minimum the amount of energy consumed (and carbon generated) by a building over its lifetime. It embodies a series of strategies such as:

- paying careful attention to build quality to ensure airtightness
- the 'superinsulation' of the building envelope
- the use of low energy consumption mechanical equipment within the building
- the careful selection of low embodied carbon materials in the building's construction.

While these strategies are all highly laudable in themselves, it should be appreciated that the choices encouraged by the concept are significantly climate-led and regionally related. The decisions as to which strategies should be adopted will vary significantly depending on the location in which the building is to be constructed. The Passivhaus approach also does not take account of occupant cultural or behavioural patterns, which can be found to affect energy consumption significantly. The approach can be spectacularly successful in the right circumstances, but its various aspects require careful consideration by the designer as to which are appropriate to pursue. It should, therefore, not be seen as a 'one size fits all' panacea. Further information is provided in *The Passivhaus Handbook* (Cotterell and Dadeby, 2012).

Large, voluminous spaces such as multi-purpose halls, which are to be used intermittently, should not be heated by convection. This will only result in a stratified layer of warm air at high level, unless the pre-heating time is long enough to fill the space from the top down. This greatly extends the heating period, and thus increases running costs. A better solution would be underfloor heating or a heated ceiling. A common misunderstanding is that heated ceilings have to be relatively low to be effective. This is not true, as the energy reaches the floor by radiation not convection, and is typically being emitted from an extended linear source, not a localised point source.

3.5.8   Heat emitters

In most modern buildings the heat source (where fuel is converted into heat) and the heat emitters (where heat is discharged into the room) are separate components. (One exception is the increasingly popular wood-burning stove.) Emitters deliver heat primarily by convection and radiation. This process varies considerably for different types of emitters, and can have a significant impact on comfort and energy consumption, as can the occupancy regime of the building (i.e. whether it will need to be heated continuously or intermittently) (Box 3.2).

3.5.9   Cooling – the absorbing and discharging of energy

The process of cooling involves the absorption and discharge of energy, in order for the building's occupants to preserve a comfortable heat balance. Energy can be absorbed and discharged, and thus cooling brought about, in two principal ways: by making use of the building's fabric (thermal mass) as an absorber and re-emitter (absorptive cooling), or by discharging air containing water vapour (and therefore energy) out of the building (ventilation cooling).

It is possible to enhance the cooling performance of energy-absorbing surfaces by running chilled water or air through them to reduce their initial temperature. Chilled beams and proprietary construction systems such as TermoDeck make use of this type of technology.

3.5.10   The nature of air, and psychrometry

When considering the internal environment in a building it is important to understand the interrelationship of several different properties of air. Air is a complicated mixture of different gases (primarily nitrogen and oxygen), vapours (primarily water) and fine particulate matter such as dust, pollen, combustion particulates, etc. Each of the constituent parts of air has its own physical properties, and responds differently to variations in volume, temperature and pressure.

The water vapour present in air contains energy, known as 'latent heat' or 'enthalpy'. In considering the amount of energy present in a mass of air, it follows that it is necessary also to know its volume, pressure and temperature. The study of these interactions is known as psychrometry. The relationships between temperature, pressure and enthalpy are published for use by designers as psychrometric charts by organisations such as the Chartered Institution of Building Services Engineers (Figure 3.2).

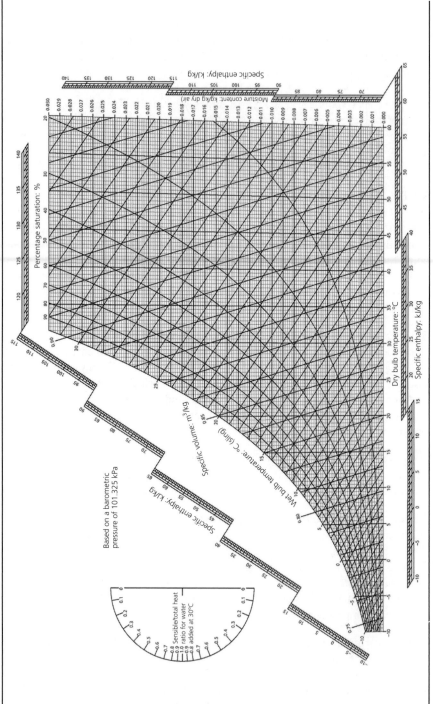

Figure 3.2 Typical PC 01 psychrometric chart for the temperature range −10°C to +60°C, published by the Chartered Institution of Building Services Engineers, showing the physical properties of air and water vapour mixtures in relation to temperature. (Courtesy of CIBSE)

It is possible by mechanical processes to manipulate the temperature, pressure and volume of a controlled mass of air, and thereby to adjust its relative humidity, water vapour content and energy content, in the interests of providing a suitably comfortable combination of air temperature and humidity within the building. These are the key processes used (with filtration to remove particulates) in air handling units (AHUs).

Broadly similar processes, involving energy transfer through the evaporation and condensation of refrigerants by compression and re-expansion, can be used to control the temperature of liquids, especially water. These are the key processes used in AHU compressive chillers.

Significant amounts of energy are needed to achieve the compression of the refrigerant and to force the air mass through the AHU, and this can have a marked influence on the overall energy consumption of the building. It is therefore advisable to reduce the building's cooling load as far as possible in the first place.

3.5.11   Thermal labyrinths and earth tubes

An energy-efficient way of reducing the building's cooling load is through the use of a thermal labyrinth or earth tube to temper the incoming air at source, as it enters the building. Incoming make-up air is led through a thermally-absorptive underground duct or structural matrix, before undergoing conditioning prior to discharge into the building. The comparatively low, steady temperature of the ground at depth serves to keep the internal surfaces of the thermal labyrinth relatively cool, allowing energy to be absorbed from the incoming air stream (Figure 3.3).

3.6.   Natural ventilation
3.6.1   Background

The provision of adequate ventilation is an essential aspect of building design. The primary functions of ventilation are to maintain a reasonable standard of internal air quality, to remove excess heat from within the building, and to provide an adequate level of physiological cooling for the building's occupants, in the interest of ensuring a suitable level of comfort.

Historically, all buildings were naturally ventilated to a greater or lesser extent, and it was only in the 20th century that mechanical systems were developed, in response to changes in the scale and nature of buildings and the functional requirements of the processes housed within them (Box 3.3).

In due course, changes in social lifestyle and the comfort expectations of occupants became a key factor in the adoption of mechanical systems. Latterly, the consideration of energy efficiency has become a driving factor, and the concept of using energy-consuming mechanical systems to maintain close control of internal environmental conditions within buildings is increasingly being questioned.

Alternative solutions that accept a less stringent level of control of internal environmental conditions are now being 'rediscovered' by building designers, and the use of natural

Figure 3.3 The Cooperative Group office building at One Angel Square, Manchester, makes use of underground thermally absorptive earth tubes to reduce the temperature of incoming ventilation air by up to 4°C at source, thus helping to minimise the cooling-system energy consumption

Box 3.3 The importance of indoor air quality

The Health and Morals of Apprentices Act was introduced in the UK in 1802. One aim of the Act was to ensure that adequate ventilation was provided in mill buildings, in order to improve the health of the industrial workers within. This followed a series of outbreaks of disease, including smallpox, typhoid fever and cholera, among workers in several East Lancashire mill towns, which were traced back in part to the conditions of close contact, poor personal hygiene and lack of fresh air that prevailed generally in the mills of the time. The Act acknowledged formally for the first time that indoor air quality (IAQ) was an issue to be taken into account in the design of industrial buildings.

Table 3.1 Differing ventilation purposes and requirements

Ventilation	Purpose	Ventilation rate
Minimal ventilation	To maintain air quality	Typical winter case: 0.75–1.5 air changes/hour
Space cooling	To vent unwanted heat	Typical summer case: 2–12 air changes/hour
Physiological cooling	To provide direct air movement to occupants	Typical summer case: 0.5–1.5 m/s air velocity

ventilation is one of these. However, it must be understood that such ventilation inevitably involves linking a building's internal environmental conditions with the prevailing external ambient conditions, whatever those may be. It is therefore not feasible to implement a natural ventilation strategy in all cases, particularly within buildings that house processes which are sensitive to variations in temperature, relative humidity or the particulate content of the ventilating air. Acoustic problems caused by noise break-in may also rule out the possibility of using natural ventilation.

3.6.2   Can mechanical ventilation be avoided?

Traditionally, building envelopes were relatively permeable. This often provided a 'fail-safe' for air quality. However, with the growing trend to make envelopes airtight in order to save energy, ventilation has to be designed more precisely. Three different purposes of ventilation can be identified (Table 3.1), and each needs to be considered separately, although their provision may be by the use of common elements.

The different purposes for ventilation make very different and sometimes conflicting demands on the building. In winter, the problem is to exchange just enough air to maintain sufficient air quality. The well-known mantra 'build tight, ventilate right' implies that the ideal is to have an airtight envelope with purpose-made controllable ventilation openings, positioned to give the best mixing and to minimise discomfort from cold draughts. Openings may be windows, or closable grilles or trickle vents. If the openings are windows only, it must be possible to set them to a very small opening area. So, while it may be possible to dispense with mechanical ventilation in some instances, this will not be possible in internal spaces, where some form of forced ventilation will be needed, and in areas that require a high throughput of air to maintain minimum indoor air-quality standards and to control odours. Kitchens and toilets, in particular, often require special attention.

3.6.3   The driving forces of natural ventilation

There are two distinctly different forces that enable natural ventilation: namely, differential pressures and differential densities.

■   Differential pressure: wind generates pressure differences across the building, and these cause air to flow into and out of openings in the building envelope (Figure 3.4).

Figure 3.4 Wind pressure distribution – airflow takes place between openings at different pressures

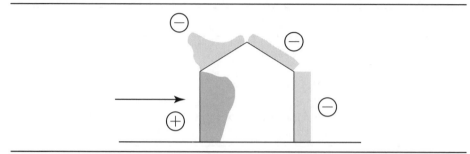

- Differential density: temperature differences between the air internally and externally, or in different locations within the building, give rise to volumes of air of different densities. This effect can be manipulated to create a vertical pressure gradient, by which the less dense air rises upward. This is known as 'buoyancy flow' or, more commonly, 'the stack effect' (Figure 3.5).

Both of these driving forces require the presence of suitable airflow paths into, within and out of the building if they are to be used effectively.

For much of the time, both these effects are present simultaneously. The problem is that both are highly variable. The overall air-change rate on a cold, windy day will be many times that on a warm, calm day, unless the openings can be made to respond to changes in the flow resistance. This will generally involve a need for some form of mechanically-operated shutter, as well as monitoring and control equipment, which will require an input of energy for its operation.

3.6.4   Winter ventilation
- Openings should be small and controllable to account for different wind strengths and temperature differences.
- Openings should be high up in the external wall of the room to encourage mixing and minimise draughts.

Figure 3.5 Temperature difference creates a pressure gradient, causing air to be drawn in through openings at the base, rise through the building, and be discharged out of openings at high level

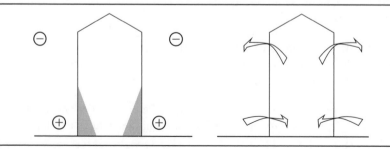

If only fixed trickle vents are provided, they have to be large enough to cope with the worst conditions (i.e. no wind and small temperature differences). This will lead to over-ventilation in conditions of significant wind and large temperature differences, and hence wasted energy. Automatic vents are now available that close as the pressure differences get greater, thereby stabilising the airflow rate. Alternatively, active ventilators controlled by a computerised building management system can be modulated in response to temperature and wind speed or indoor air quality.

3.6.5   Summer ventilation

■ Openings should be large and easily controllable (good access to handles, stays, locks, etc.).
■ Openings should be well distributed vertically and horizontally to encourage flow between parts of the façade at different pressures (Figures 3.6 and 3.7).

Figure 3.6 Wind generates complex pressure distributions on buildings, particularly in urban environments. This assists ventilation, provided that openings are well distributed and flow paths within the building are available

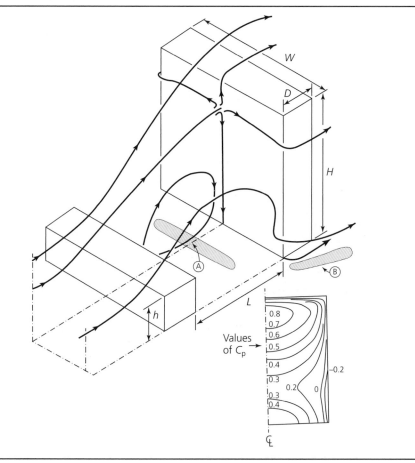

Figure 3.7 Ventilation is improved when openings are well distributed vertically and horizontally. This is because airflow is driven by the differential pressure between adjacent openings. It also leads to a better distribution of air within the room.

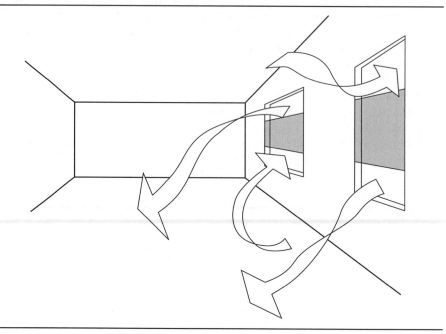

■ Consideration must be given to the way in which the incoming air will affect the occupants.

■ Shading devices must not block summer ventilation openings.

■ Openings may require special design features to reduce transmitted noise.

In both cases, consideration must be given to the distribution of fresh air within the space. Different distributions of openings allow different depths of floor plan to be naturally ventilated. Table 3.2 gives rules of thumb for the proportioning of rooms.

Note that the depth of effective ventilation is dependent on the floor-to-ceiling height. The removal of a suspended ceiling may represent a refurbishment opportunity for

Table 3.2 Depth of effective natural ventilation in rooms from side openings

	Single-sided		Double-sided (cross-ventilation)
	Single opening	Multiple openings well distributed vertically and horizontally	
Depth of room in units of floor to ceiling height h	$2h$	$3h$	$6h$

Figure 3.8 Stacks can ventilate a deep-plan building; wind and buoyancy forces both create differential pressure, causing fresh air to flow in from the perimeter

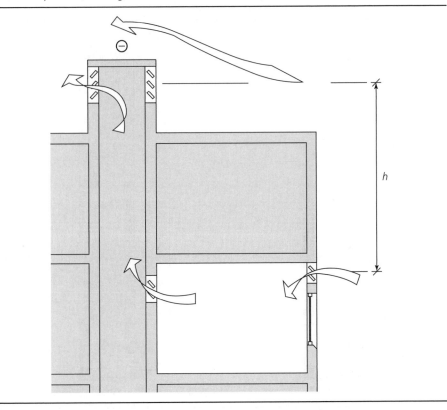

improving natural ventilation (and daylighting), provided other problems such as servicing and acoustics can be solved.

3.6.6 Advanced ventilation techniques

So far the guidance given has considered ventilation provided by openings in the façades of the building, these usually doubling up as windows. This imposes some limitations on plan depth and layout. To get sufficient quantities of air in and out of an existing deep-plan space, ducts and/or chimneys are likely to be needed (Figure 3.8).

3.6.7 Vertical ducts and solar chimneys

Large vertical ducts can generate larger airflows by using the stack effect than can be obtained in a single room by side openings, because of their greater height. Furthermore, when the wind does blow, an area of negative pressure area is typically created across the top of the building, and thus the wind-driven and stack-driven flows complement each other. In order to keep flow resistance low, the duct cross-sectional area has to be relatively large, typically 2–5% of the floor area being served. Note that it is also necessary to provide a similar area of inlet into the space being ventilated, to allow 'make-up air' to

Figure 3.9 Solar chimneys use solar gains to heat the air column. However, the driving force is dependent on the height of the warm-air column and the average temperature difference. It follows that there is no benefit in heating up the air at the top only

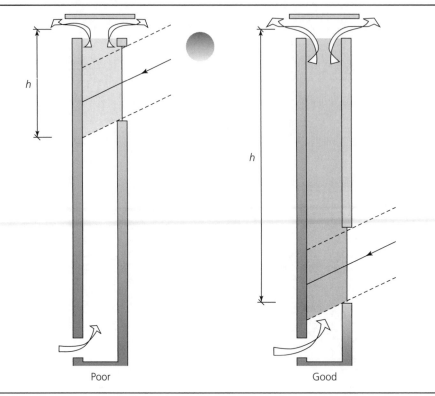

Poor   Good

enter. This may be difficult to provide in an existing building, although use of existing light wells or redundant lift shafts might allow a workable solution.

The performance of the stack can be enhanced by heating the air within it through the use of solar energy. Ideally, the air at the bottom of the stack is heated, as it is the relatively low density (and hence greater buoyancy) of the warmer air that drives the flow up and out of the stack. These elements are often referred to as 'solar chimneys' (Figure 3.9). They may not work well in cold and sunless weather conditions, when the poorly insulated glass surfaces allow the air in the chimney to become cooler than the surrounding air in the building. It thus becomes more dense, sinks downward, and generates a reverse flow. The need to allow solar radiation access to the stack to heat the air inside means that solar chimneys cannot be located towards the centre of deep plan buildings.

3.6.8   Night-time ventilation to provide cooling
The thermal mass of buildings can be used to absorb heat gains through the day, thus reducing peak internal temperatures (and, consequently, the cooling load). However, the absorbed heat has to be discharged from the building at some time, and this is done most

Figure 3.10 Night-time ventilation can reduce daytime temperatures by as much as 4°C. However, it only works where there is thermal mass available internally, and high rates of night-time ventilation

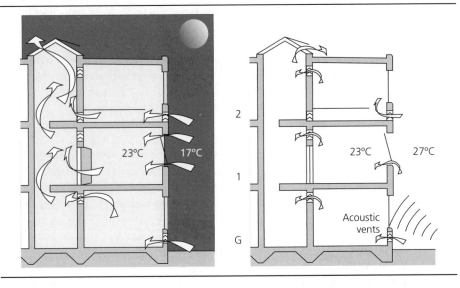

efficiently by maximising the ventilation rate at night, when the external ambient air temperature is lower (Figure 3.10). When employed successfully, this technique can reduce the daytime temperature in the building to a level 3–4°C lower than the peak outdoor temperature. In this case it pays to reduce the ventilation rate during the day when (and if) the outdoor temperature is above the indoor temperature.

When implementing night-time ventilation strategies, the following aspects need to be considered:

- Provide openings that can be left open at night, yet maintain security.
- Consider how large volumes of air will flow through the building, from room to room and from floor to floor.
- Thermal mass is only effective if it is exposed within the occupied space, and accessible to the cool ventilating airstream.
- In lightweight buildings, consider adding thermal mass (e.g. thermally absorptive ceilings or partitioning) or even phase-change materials.

3.6.9   Hybrid natural/mechanical systems

Natural and mechanical ventilation need not be mutually exclusive, even though certain spaces in a naturally ventilated building, such as toilets or kitchens, may still need to be mechanically ventilated if they have no direct access to outside air, or have high ventilation demands.

Significant energy can be wasted if ventilating unoccupied or lightly occupied spaces. 'Demand control' can be used, where fans are only run when the air quality (defined by carbon dioxide (CO_2) levels) is deemed unsatisfactory. Reliable, inexpensive CO_2

detecting controls are available. If the wind and buoyancy forces are too weak, low-pressure fans can be used to supplement the natural airflow in ducts and chimneys. The fans also need to be activated by a control system, which detects a reduction in airflow, or a fall in air quality. There are arguments for and against hybrid systems.

- *Argument for*: passive systems have to be oversized in order to cope with worst-case scenarios. By accepting mechanical intervention with appropriate controls, an optimum balance between energy efficiency and comfort can be struck.
- *Argument against*: a hybrid system necessitates the capital expense and maintenance of two systems.

One technique where a small amount of mechanical power can be used with great effect is the use of ceiling or desk fans. These devices provide air movement but not fresh air. The air movement can make a reduction in the effective temperature as perceived by the occupants, of as much as 3°C. However, although they improve thermal comfort, they do not improve air quality.

3.7. Daylighting
3.7.1 Background

The incorporation of natural daylight into buildings is a complex area of design. Good-quality daylighting design can be both functional and emotional in its effects. From the viewpoint of energy, lighting is frequently a major consumer of electricity in buildings, so the appropriate effective use of natural daylight can significantly reduce power consumption. It can also have a demonstrably beneficial effect on building occupants, both physiologically and psychologically. However, allowing daylight into buildings is not possible without fenestration. Daylight is solar energy; the admission of daylight involves solar gains, and the admission of significant levels of daylight can substantially influence a building's energy balance and increase cooling loads. Daylighting can, therefore, be seen as something of a double-edged sword.

Several important basic aspects of daylighting are discussed below, but no calculation methods are given, as this is a highly specialised area of design, and detailed technical information can be found from other sources (see, for example, Baker and Steemers, 2002).

3.7.2 Energy

The proportion of electrical power used by lighting in buildings depends on several factors, including the building's internal layout and façade design, the global location and season, and, not least, the building usage. Broadly speaking, estimates may range from around 5% to 50%. The careful incorporation of daylighting can, therefore, have a marked effect on the building's carbon footprint.

In cold climates, the use of large areas of façade glazing to facilitate daylighting can also allow the beneficial capture of solar gains, thereby providing a secondary benefit of reducing peak heating demand, although the reliability of the sun to provide energy in this way clearly cannot be guaranteed in most climates.

Conversely, in warmer climates, added solar gain can be a curse, leading to rapid over-heating, and dramatic increases in cooling load, operational energy requirements and operating costs. It is therefore important that very early in the design process the drivers for the incorporation of daylighting are reviewed and a check made of what the anticipated benefits may be. Allowing for the effective control of incident sunlight is critical, and needs to be considered from the outset.

3.7.3 Glazed façades

The holy grail of the invisible, see-through façade has been pursued by building designers for many years. The metaphorical and literal property of transparency appears to have an irresistible allure, enabling views both out and in, and thereby adding a sense of clarity and understanding of the building's function and purpose.

Glazed façades also constitute the most difficult area of the envelope to insulate, and constitute the area most susceptible to both energy loss and solar gain, with U values potentially between five and ten times greater than the adjacent, non-glazed façade. So, if daylighting is seen as a desirable attribute of the building, requiring large areas of glazing to be provided, seasonal variations in heating and cooling demand must be taken into consideration and modelled carefully at the early design stage.

The word 'transparent' does not necessarily imply the use of glass. The development of ethylene tetrafluoroethylene (ETFE) fabrics and inflated pillow technology has provided the designer with an attractive alternative option to admitting natural daylight without suffering the penalty of excessive energy loss. The ETFE pillow walls to the main atrium space at South East Essex College, Southend, are an interesting example of the application of this approach (Figure 3.11).

3.7.4 Direct and indirect natural light

Admitting high levels of natural light into buildings is not always beneficial. The intensity of direct incident sunlight in the UK is of the order of 40 000–80 000 lux, depending on the season, whereas a typical target working-plane illuminance in an office building is around 500 lux. This huge difference in intensity ('glare') can give rise to severe discomfort. It is therefore important to be able to control incident sunlight to within acceptable levels. The orientation, size and positioning of windows plays a vital part in this, as does the use of either fixed or operable shading.

A useful technique is to allow daylight into the building after having reflected it off other surfaces first. High-intensity direct incident light can then be excluded, and the level of admitted light can be regulated to some extent by adjusting the reflectance of the surfaces on which it impinges. The use of light shelves, high-albedo external surfaces adjacent to the façade, and water-filled reflecting pools are all practical manifestations of this approach (Figure 3.12).

3.7.5 Room proportions and fenestration ratio

It is important to consider two sets of dimensional information if it is intended to make use of natural daylight to light a room. The first of these is the ratio of the window head

61

Figure 3.11 At South East Essex College, Southend, a light, airy feel is achieved with acceptable energy losses by using an ETFE pillow façade in the main atrium space. (Courtesy of Tristram Hope)

height to the depth of room away from the window wall. As a rough guide, if this ratio exceeds 2.5 the variation in illuminance across the depth of room is likely to be excessive: those adjacent to the window may find it too bright, while those to the inside of the room are likely to find it too gloomy (perhaps 10% or less of the window-side illuminance) for normal daily tasks. The key dimension here is the height from the floor to the window head, which should be maximised.

The second is the proportion of the window wall area that is glazed, or made translucent. It is unlikely that a room with less that 40% of glazing in the façade will allow for effective daylighting, although this depends to some extent on the use to which the room will be put. Clearly, this can have a marked impact on the building's overall external appearance, and so must be an early design consideration.

3.7.6   Where not to use natural daylight
Not all spaces are suitable for natural daylighting. Any process that requires a constant temperature level or a constant or low level of light is likely to be adversely affected by natural daylighting, as the external levels of illuminance can vary greatly. Spaces that house temperature-sensitive manufacturing processes or computer visual-display equipment are commonly-encountered examples.

Figure 3.12 Tempered, indirect daylight used to spectacular effect in the Bishop Edward King Chapel, Oxfordshire. (Courtesy of Balazs Bicsak)

3.7.7   Daylight linking

Daylighting by itself is unlikely to provide a satisfactory solution; almost invariably, artificial lighting is used to supplement natural daylight. Where this is the case, consideration should be given to the 'daylight linking' of luminaires within the space being lit, using light sensors. The sensors detect light levels within the space, and can be arranged progressively to switch the lights on or off as required, to maintain a more even distribution of light across the full floor area, thus avoiding the provision of excessive light levels adjacent to the window wall, with the consequent waste of energy.

Where this technique is used, it makes sense to arrange the circuitry of the luminaires in lines parallel to the window wall (Figure 3.13) so that progressive banks of luminaires can

Figure 3.13 A typical classroom at Ifield Community College, Crawley, showing the use of natural daylight in conjunction with daylight-linked artificial lighting, with luminaires arranged in ranks parallel to the window wall to facilitate phased switching

be switched on, starting with those towards the inner (darker) side of the room as light levels start to fall, and vice versa, starting with those adjacent to the window wall, which are switched off first as light levels pick up.

What can engineers do?

- Adopt a holistic, integrated, multidisciplinary strategy towards building design, as opposed to a segregated single-discipline, 'silo' approach.
- Develop an understanding of the fundamental principles of building physics, in order to be able to distinguish between inherently feasible and non-feasible solutions, to be able to 'leap-frog' the traditional evolutionary process of technology development, and to keep pace with the rapidly changing performance requirements put on today's built environment.
- Design to facilitate adaptive behaviour by providing adaptive opportunity in buildings, including ergonomic, intuitive local control systems to allow individual action to modify the internal environment in the building.

- In the case of buildings that use mixed-mode or passive systems, encourage building owners, managers or operators to provide periodic teaching sessions for the building occupants, to ensure that they are aware of any interventions that may be needed to keep the internal environment comfortable, while keeping energy consumption to a minimum.
- Encourage the development of natural feedback loops; design main building control systems to default to a 'caretaker' mode; recognise the potential benefits of adopting a comfort-driven dress code.
- Adopt the general policy of minimising the use of energy and natural resources, coupled with a strategy of adaptation to changing circumstances.
- Take account of orientation, solar geometry and sun path when positioning the building on site, and when making key decisions about the size and position of areas of glazed façade.
- Understand the interrelationship of different properties of air, in particular the response of its different constituents to variations in volume, temperature and pressure.
- Facilitate the use of high-impact, locally-controlled plant such as desk fans and task lighting, to allow the rapid, localised modification of the user environment.
- Encourage the use of natural daylighting where appropriate, and recognise the need for localised control and the potential problems associated with solar gain. Make provision for daylight linking of artificial lighting to minimise energy wastage.

REFERENCES

Baker N and Steemers K (2002) *Daylight Design of Buildings: A Handbook for Architects and Engineers*. Routledge, Oxford, UK.

Cotterell J and Dadeby A (2012) *The Passivhaus Handbook*. Green Books, Cambridge, UK.

Maslow AH (1943) *A Theory of Human Motivation. Psychological Review* **50**: 370–396.

FURTHER READING

Baker N (2009) *Sustainable Refurbishment in Non-Domestic Buildings*. Earthscan, London, UK.

Baker N and Steemers K (1999) *Energy and Environment in Architecture – A Technical Design Guide*. Taylor & Francis, London, UK.

CIBSE (Chartered Institution of Building Services Engineers) (2014) Various journals and guides. Available at www.cibse.org (accessed 12/11/2014).

Humphreys MA (1976) *Field Studies of Thermal Comfort Compared and Applied*. Building Research Establishment, London, UK, Current Paper 76/75.

Szokolay SV (2014) *Introduction to Architectural Science: The Basis of Sustainable Design*, 3rd edn. Routledge, Oxford, UK.

Tregenza P and Loe D (1998) *The Design of Lighting*. E & FN Spon, London, UK.

Tregenza P and Wilson M (2011) *Daylighting: Architecture and Lighting Design*. Routledge, Oxford, UK.

Sustainable Infrastructure: Sustainable Buildings
ISBN 978-0-7277-5806-4

ICE Publishing: All rights reserved
http://dx.doi.org/10.1680/sisb.58064.067

Chapter 4
Planning for in-use to end-of-life

Elisabeth Green

By failing to prepare, you are preparing to fail.

Benjamin Franklin

4.1. Introduction

The way in which the built environment industry considers buildings in use is changing. The 'in-use to end-of-life' aspect of a building's life cycle is often viewed as its operation and maintenance, and is thus considered to be about driving a reduction in energy consumption through the efficient commissioning of mechanical and electrical systems, decommissioning, and the way in which a building is structurally deconstructed or reused. This chapter focuses on *planning for* 'in-use to end-of-life' (if you are an engineer working on a building already in use see Ainger and Fenner, 2014: Ch. 10).

However, alternative approaches to this traditional 'operation and maintenance' view are also discussed by considering how occupants can directly affect a building in-use and the way in which the building's environmental performance is affected by its wider context. Ways of creating a more sustainable building are explored by introducing the following concepts:

- the way in which people interact with their urban environment
- the way in which the sustainability performance of buildings is affected by the wider context of transportation, water and waste
- energy reduction through designing to encourage adaptive occupant behaviour during post-occupancy evaluation.

4.2. How the sustainability performance of buildings is affected by their wider urban context

4.2.1 Introduction

Built environments are responsible for providing space for people to inhabit, work and socialise. The latest recorded data show that in 2010 the world's population was close to 7 billion people, that there are 21 megacities (a city with a population of over 20 million people) and that global carbon emissions need to be reduced by 80% by 2050 for future generations to experience only a 1–2°C average increase in climatic temperature (United Nations, 2012). The last two centuries have seen an increase in population migration from rural to urban areas driven by the need for employment during economic recession, the increased cost of living, a need for proximity to amenities, as well as population growth through an improvement in social and health care, which has led to the United Nations forecasting that 'today's urban population of 3.2 billion will rise to nearly 5 billion by 2030, when three out of five people will live in cities' (Lewis, 2007). Many of the existing megacities struggle to manage traffic congestion, air pollution, crime, power cuts, sanitation, social care and unemployment. So what can be done to improve megacity life?

By understanding how people live in their individual 'space' and by understanding occupant behaviour, designers can start to improve schemes at an early stage to create more adaptable built environments across cities. It should also be understood that the thinking outlined in this chapter is a growing area of research being led by radical thinkers. As the construction industry starts to understand the occupant–building relationship, design choices, processes and strategy will inevitably change. To put this thinking into context, this section introduces the characteristics of this relationship between people and urban space, and how behaviour is influenced by transport, water and waste management on a city scale.

4.2.2   Changing behaviour of people within their urban space

Understanding the relationship outlined above can influence decisions that can lead to social, economic and environmental change. An example of social change is the 'zero tolerance' policies introduced by New York City's Mayor Guiliani in the mid-1990s, and further developed by Mayor Bloomberg, to improve social behaviour in the city. The zero-tolerance approach to crime sent out a clear message to people that their behaviour must change. And behaviours did change, making it safer to use the subways and walk along the sidewalks of the city without fear. Another example of behavioural change occurred in Japan in 1997 with the introduction of the Law for the Promotion of Sorted Collection and Recycling of Containers and Packaging (Government of Japan, 1997). This has influenced the behaviour of the population, leading to an increase in the country's plastic recycling rate to 77% – which is over twice that of the UK, and also greater than the rate of 29% in the USA (PWMI, 2012). Japan has also enforced further recycling laws, and is benefiting from the results as companies like Panasonic are changing their product and packaging designs to reduce waste and save money.

These examples show that people's behaviour can be changed if the right drivers are in place. A prime example of a simple change in behaviour resulting in the reduction of significant environmental impact was the banning of chlorofluorocarbons (CFCs) in the 1990s, which stopped the hole in the ozone layer increasing in size and therefore had a global benefit. A design team should consider its role in ensuring behavioural change and the skills required by the team to drive the intended change (Box 4.1).

An example of effective local change through community involvement is the King's Cross Skip Garden, London, which is run by local children and community volunteers, and hosts a pop-up on-site café (King's Cross Skip Garden, 2014).

In the following sections, examples are given of how consideration of the urban environment can drive behavioural change in the areas of transportation, water and waste in

Box 4.1 What can design teams do to encourage behavioural change?

■ Hold workshops with building users or community groups.
■ Maintain contact with these groups throughout the design process to create a sense of ownership.
■ Undertake a needs analysis based on site visits, and research the local social, economic, cultural and environmental profiles.
■ Design into the solution an educational legacy that allows users to interact with their environment.
■ Apply a sustainability hierarchy (e.g. Ainger and Fenner, 2014: Ch. 10, Figure 2.3).

order to facilitate sustainable solutions. In each case the design team will need to adopt an approach which ensures that their building fits into the wider urban context, otherwise it will not achieve its optimum performance under social, environmental and economic criteria.

4.2.2.1 Transportation

The way in which a development is planned will influence travel and transport decisions through its subsequent population and the population around it. The location of key trip origin and destination points and accessibility to different transport modes will affect the journey time, chosen method of transport, and subsequent traffic density. Although these factors might be considered as separate parameters, they can all lead to the resultant reduction in the levels of vehicle emissions, therefore improving environmental performance.

For example, Mexico City suffers from poor air quality, mostly due to vehicle emissions, but in 2013 the city won the Sustainable Transport Award for its planning-related actions to reduce congestion. The city extended its Metrobus system though narrow, congested streets, rebuilt public parks and plazas, expanded bike lanes and bicycle sharing (Figure 4.1), and created more pedestrianised streets.

Cumulatively, these interventions led to less traffic congestion and pollution, and people being able to move more easily around the city, using more sustainable modes of travel. Mexico City is just the latest urban area that has recognised the value of integrating sustainable transport planning into its urban master plan. Fifty years ago, the Brazilian city of Curitiba acted as a trend setter, providing cheap, sustainable access across the city by way of its integrated bus rapid-transit system, which built on and enhanced the existing urban fabric.

Other cities have continued to follow suit and looked to incorporate such elements of transport planning into the urban realm. The cycle-friendly cities of The Netherlands are often cited as examples where the road hierarchy puts pedestrians and cyclists at the top. This integrated aim of public transport is easier to achieve when people live and work across urban areas than when creating separate living/working/leisure zones that require frequent, lengthy (often car-based) trips. In the United Arab Emirates, neighbourhood

Figure 4.1 Mexico city's EcoBici (bike sharing scheme) in practice. (Courtesy of Luisus Rasilvi)

developments with courtyards (*fareej*) often use shaded walkways consisting of *sikka* (narrow streets shaded by surrounding buildings that connect community facilities and residences together) and *barahat* (partially shaded, intimate public spaces). In some neighbourhood developments in Al Ain and Al Sila'a, each villa has also been provided with cycle parking. The aim is to positively encourage walking and cycling as the main modes of transport between villas and local services. Further examples of how transport planning for an individual development can be very interesting and diverse because it encompasses not only how the buildings are initially designed but how they can be adapted over time are given in Box 4.2.

Although alternative fuels, more efficient cars, bicycles and walking all have a significant role to play in sustainable transport, sometimes it is more effective to reduce the requirement to travel by looking at the urban environment itself; for example, by designing multifunctional urban centres where people can live, work and play all in one area. In the UK, 21% of CO_2 emissions relate to travel, most of which is between home–work–leisure buildings.

4.2.2.2 Water
In developed countries, effective sewerage and potable water infrastructure is taken for granted. This is evidenced by the fact that potable-quality water is even used to flush toilets. A simple sustainable decision could be taken to use grey-water or rainwater recycling systems instead. In developing countries access to potable water is restricted either by geography or by resource; sometimes potable water is created through

---

Box 4.2 Examples of planning for transportation

- **Microsoft**. From its major offices in London and Dubai to smaller local offices such as the one in Abu Dhabi, the company's approach to building use and design is obvious. Flexible working is encouraged so that people only use the office – and travel – when needed. The impact of this is that space can either be reduced or given over to better uses, such as customer-facing activity.
- **Environment Abu Dhabi**. The company is adopting an integrated fleet-management approach which aims to more efficiently use, and hence reduce, the current operational fleet. Through this, methods of working will be impacted. For example, there will be less unnecessary travel for business meetings and more optimisation of technology.
- **London's 2012 Olympic Park**. Legacy planning has put walking and cycling at the top of the transport hierarchy. Residential accommodation is planned next door to local business centres to encourage sustainable movement by decreasing long-distance car-based travel.

---

desalination, which uses a significant amount of energy to produce water for consumption. Other places rely on handpumps for water supply. In the World Wide Fund for Nature (WWF) report *Big Cities, Big Water, Big Challenges* (Kraljevic, 2011) it is identified that 'the main threats to urban water are water scarcity, decreasing water quality and pollution, water over-use and associated salt-water intrusion, in addition to infrastructural, institutional, and social problems'. Thus, all countries have a responsibility to reduce demand through leak management and efficiency measures, and by investing in new civil infrastructure that can manage storm water events and cope with waste water from the increasing population in cities. Other options include sharing water between countries through a 'global water grid' (Box 4.3).

Due to a change in rainfall patterns since 1990, water-supply problems and a rising population, Mexico City has taken steps with its Plan Verde to preserve water resources. Plan Verde cd de Mexico is seeking to ensure an adequate future water supply by reducing residential water use and network losses, recharging the local aquifer, providing water metering for all users, and increasing the volume of wastewater to be treated for re-use.

Reducing water demand in buildings is a mandatory requirement in most eco-rating systems for buildings. Typical responses are through improving the sanitary fixtures and fittings, reducing the pressure of the system, and providing rainwater storage, grey-water or domestic black-water filtration systems. A design caution to note for post-occupancy behaviour is that, if the building occupier is not aware that these systems are in use, then they are unlikely to be maintained, and thus will become inefficient and, eventually, stop working.

4.2.2.3 Waste
Globally, waste accounts for 3% of greenhouse gas emissions (Arup and C40 Cities Climate Leadership Group, 2011). Cities create large volumes of waste, much of which

> **Box 4.3** The possibility of a global water grid?
>
> From a utilisation perspective it could be argued that one way to share water between countries having excess water and those experiencing drought could be through a 'global water grid', although the reality of creating one would have economic, environmental and cultural challenges. For example, it is very energy intensive and expensive to move water over long distances where gravity cannot assist the flow. So, for now, in developed countries, the focus is on educating the population to be responsible consumers, and working to provide remote communities with access to drinking water and introducing pico hydro systems (Figure 4.2). The cost of long-distance transport of water can be compared with the cost of local desalination in order to gauge the relative cost-effectiveness.
>
> **Figure 4.2** A pico hydro system in use. (Courtesy of National Collegiate Inventors and Innovators Alliance)
>
>

goes to landfill. For example, New York City residents alone generate 12 000 tons of waste a day (Grow NYC, 2014). When waste is generated in this quantity, the embodied carbon in the waste products becomes significant, to the point where it is commercially viable for it to be reused or recycled. Waste can only be recycled or reused commercially if it is managed at source or diverted from landfill. In Delhi, research carried out by UN Habitat states that 'informal recyclers handle 27% of the waste generated ... if they were to disappear, the city would have to pay its contractors to collect and dispose of an additional 1800 tonnes of waste every day' (UN Habitat, 2010). Dealing with waste at source tends to be managed informally in developing countries because of the lack of infrastructure to support such activity. Ironically, waste can produce real value for some individuals, although of course with serious attendant health and safety issues.

The embodied carbon in a building's fabric can also be important to a building's 'in-use and end-of-life' by consideration of the flexibility of space. Where buildings are designed to be totally demountable, reused and recycled, the impact of this embodied carbon can be reduced. The London 2012 Olympics Aquatics Centre is a good example of this, where the demountable 'wings' used for the extra spectators are being shipped to Brazil to be reused in the 2016 Olympics in Rio de Janerio.

4.2.3 What is the collective impact of adapting occupant behaviour on a city scale?

People interact with water, waste and transport every day, and by considering this on a global scale opportunities for cross-country problem solving can arise, as illustrated by China's South–North Water Diversion Project which, when completed in around 40 years, will be one of the world's biggest feats of engineering, taking water from the south of the country to the arid northern region, which suffers from water shortages (Duggan, 2013).

Megacities have recognised the need to facilitate change, and in 2005 the C40 global megacities network was established by, the then London Mayor, Ken Livingstone. The network is 'committed to implementing meaningful and sustainable climate-related actions locally that will help address climate change globally' (C40 Cities Climate Leadership Group, 2013a), and provides cities with a forum for discussing their individual characteristics and vulnerability, while collaborating with other cities to solve shared problems. C40 currently represents 58 cities over five continents which collectively generate 18% of global gross domestic product (GDP), are home to one in 12 people worldwide, consume two-thirds of the world's energy and produce 70% of the world's CO_2 emissions. C40 has identified eight factors that will improve opportunities to combat climate change: waste, ports, buildings, transport, energy, renewables, lighting and water (Arup and C40 Cities Climate Leadership Group, 2011). Table 4.1 summarises the main problems and potential solutions for each of these factors.

All the examples in Table 4.1 show that the impacts of the wider urban context need to be considered during master-planning stages. In addition, an individual building needs to fit within its wider urban context in order to function as a sustainable solution, because it is often difficult for individual projects to have a cumulative beneficial effect in the absence of an urban development plan or digital city model. In the next section, the relationship between an individual building (or a small group of buildings) and the wider urban context is discussed in terms of energy usage.

4.3. Energy reduction through designing to encourage adaptive behaviour

4.3.1 Introduction

The global built environment is responsible for perhaps 30–40% of greenhouse gas emissions (UKGBC, 2013: reference to United Nations Environment Programme 2012). A large proportion of these emissions is likely due to the direct operational and in-use impacts of buildings, and therefore these factors play a pivotal role in the creation or prevention of a sustainable future. The office or home is located within an urban environment, which creates a relationship between building and lifestyle: is the occupant

Table 4.1 Summary of the main problems and potential solutions for C40 climate change factors (C40 Cities Climate Leadership Group, 2013b)

Factor	Problem	Solution	Related case study
Waste	City waste enclosed in landfills releases large amounts of methane gas as it decomposes. Methane is a greenhouse gas even more potent and toxic than CO_2.	The methane is captured and burned as a fuel to create electricity, providing electrical power and offsetting both fossil-fuel consumption and methane emission.	Toronto Sao Paulo
Port	Berthed ships typically generate energy using their own diesel auxiliary engines, generating large amounts of noxious exhaust and noise.	When docked, ships can plug into an onshore power supply, which provides cleaner energy from renewable resources.	Gothenburg
Building	Rooftops provide a huge amount of untapped space, for green space as well as sunlight capture.	Photovoltaic solar panel systems can provide energy for government buildings, while simultaneously shading them from the sun, reducing the electrical draw for air-conditioning systems and extending the life of the roof.	San Francisco
Transport	Bicycling is a zero-emission form of transport, and well suited to city life. However, a lack of cycling infrastructure in most cities discourages citizens from cycling.	A bicycle-hire service that is integrated into the city's public transport systems offers low-cost cycle hire at public transport nodes across the city, encouraging uptake of cycling and reducing emissions and congestion on the roads.	Paris Barcelona
Energy	Burning fossil fuels for energy creates many millions of tonnes of CO_2 emissions per year.	Renewable-energy sources, such as wind and solar power, reduce reliance on fossil fuels and related emissions. Also, by requiring energy companies to regularly publish data on their CO_2 emissions, some cities are driving competition for renewable-sourced energy over fossil-fuel energy.	Tokyo

Table 4.1 Continued

Factor	Problem	Solution	Related case study
Renewables	Traditional water heaters consume large amounts of natural gas or electricity to heat water from the base to the desired temperature.	Solar hot water heaters use the sun's thermal energy to preheat household water and help reduce energy consumption used by traditional water heaters.	Barcelona
Lighting	Standard incandescent bulbs used in roadside lighting are inefficient, and burn out quickly, causing delays in traffic flow during bulb replacement.	LED modules are brighter, last up to 15 times longer, and reduce energy consumption by up to 88%, which means fewer burnouts and safer traffic.	Chicago Hong Kong Portland
Water	Leaking pipes and unmonitored pressure are the largest causes of wasteful water-system inefficiencies.	The use of data loggers to monitor water pressure helps to detect, and thus fix, leaks quickly, and reduce pressure to low-flow areas.	Emfuleni Fortaleza Tokyo

able to walk to work, go to a local shop, etc., thus reducing emissions, and saving money to be able to afford to pay his or her energy and water bills?

To reduce potential future living costs, a building's designers and users have a responsibility to reduce energy consumption, but a responsible user cannot be resource efficient unless the building he or she lives or works in has also been designed to be resource efficient too. In the next section, we explore how buildings and people interact to influence 'in-use to end-of-life' through the consideration of the Exhibition Estate (London), Soft Landings and 2 St Pauls Place (Sheffield). In Section 4.3.3 we show that such behaviour change can be one of the largest contributing factors in reducing carbon footprints by considering energy consumption.

4.3.2 Consideration of occupant behaviour to share energy loads
Understanding the impact of developments within the urban context can help to reduce infrastructure loads, sourcing of materials, operational energy use and embodied carbon. By thinking on this broader scale, low-carbon technologies will work more effectively because operational energy has been minimised through adopting adaptive design. This concept is illustrated through the 1851 Group technology study Box 4.4.

4.3.3 Designing for occupancy
Once the principles of how to design a sustainable building are understood, to achieve its full potential the building user must be able to operate and live in the building in the way that was intended.

Box 4.4 The 1851 Group technology study

The 1851 Group is formed of the institutions that make up the South Kensington cultural and academic estate in London. This includes many iconic buildings from the Victorian era, but also modern buildings with energy-intensive uses as part of the Imperial College campus, as well as new additions to local museums. The buildings have widely varying demand profiles for heating, cooling and power, both over a 24-hour period and seasonally.

To understand this approach in detail one sometimes has to take a step back from the technology being used and consider simply how the buildings are used by their occupants. Imperial College is occupied 24 hours a day in some areas. The buildings contain science equipment, students using computers and lecture halls. They all generate excess heat when in use, and therefore high mechanical cooling loads, in some cases for most of the year. In contrast, the Albert Hall (an iconic venue used intermittently for concerts and celebration events with a large volume, see Figure 4.3) requires significant heating for occupants to be comfortable in winter, and similarly significant cooling in summer. Taking the buildings that produce excess heat and transferring this energy to buildings that need heating reduces the amount of energy required to heat and cool individual buildings – although this is only possible at certain times of the year. The 1851 Group technology study also looks at how inter-seasonal heat storage using an underground aquifer can further enhance the balancing of heating and cooling demands over a year. The study has reintroduced a community-scale approach to energy that can be successively adapted to connect other buildings into the network on a phased basis. In doing so, the economies of scale increase the benefit to all involved, and inefficiencies in the energy network infrastructure are reduced, leading to a potential reduction in carbon emissions of 30% (NCE, 2012).

Figure 4.3 This image shows how a historic (or iconic) building such as the Albert Hall uses lighting to create a sense of place at night. However, in doing so it consumes electricity and adds to the building's energy load, which is required to be offset in the 1851 Group study. (Courtesy of Carbon Visuals)

Understanding how post-occupancy behaviour affects a building's in-use performance traditionally starts with the handover of the building. Evidence has shown that handing over a sustainable building to an occupant who is unfamiliar with the way the building works will lead to the building underperforming. Some designers consider that one approach to improving performance is for all projects to involve stakeholder consultation (i.e. the end user is involved throughout the design process, and therefore develops a better understanding of the building's operational processes, from the project's outset to beyond its handover). To this end, BSRIA has developed a set of guidelines called 'Soft Landings' (UBT and BSRIA, 2009), targeted at making buildings work as intended and at reducing operational and energy costs.

4.3.3.1 Soft Landings

The Soft Landings Framework (UBT and BSRIA, 2009) has been developed to improve how designers, occupiers and constructors account for a building in use in order to improve how it functions and delivers against its expected performance. Soft Landings was created because modern energy-efficient buildings have become increasingly reliant on complex technology that needs careful commissioning, monitoring, maintenance and feedback by occupiers who understand how to undertake these tasks. The developed framework starts by 'raising awareness of performance in use in the early stages of briefing and feasibility', and then 'helps to set realistic targets and assigns responsibilities' (UBT and BSRIA, 2009: p. 11) and continues to outline actions through the design process and beyond for 3 years after building handover.

The five stages of Soft Landings (taken directly from the Soft Landings Framework (UBT and BSRIA, 2009)) are outlined below:

- *Stage 1: Inception and briefing*. The goals in this stage are to clarify the duties of members of the client, design and building teams during critical stages, and to help set and manage expectations for performance in use.
- *Stage 2: Design development and review (including specification and construction)*. This follows normal construction processes but with greater attention to applying procedures established in the briefing stage, reviewing the likely performance against the original expectations, and achieving specific outcomes.
- *Stage 3: Pre-handover*. This process is done with greater involvement of designers, builders, operators, and commissioning and controls specialists, in order to strengthen the operational readiness of the building.
- *Stage 4: Initial aftercare*: During the users' settling-in period, with a resident representative or team on site to help pass on knowledge, this stage involves responding to queries and reacting to problems.
- *Stage 5: Aftercare in years 1 to 3 after handover*. The periodic monitoring and review of building performance.

Post-occupancy evaluation has demonstrated that successful outcomes require collaboration between the client, the occupier and the design team, together with commitment beyond the design team's normal contractual remit to make the building design a real success. Soft Landings is discussed in more detail in Section 7.12 in Chapter 7.

4.3.3.2 Energy efficiency through informed occupants

Assuming that the building design has employed the principles of passive design, a successful energy-efficient building will need to contain 'easy to use' energy-efficient systems and appliances, and an occupier who is able to effectively use the technology provided. For example, low-energy buildings in a temperate or cold climate can use systems such as air source heat pumps (which extract energy from the air to provide thermal energy to provide heat) and heat-recovery systems to reduce energy consumption further. In addition, electrical energy consumption can be reduced through the use of daylight sensors, energy-efficient lighting and lighting controls, such as occupancy sensors. Not all these technologies might initially be known to the building user, and will require their engagement so that the relevant functionality is used in practice. Thus, post-occupancy evaluation and support must be provided in order to realise improved performance benefits.

Demand for energy efficiency is increasing because there is more political pressure to introduce energy-efficiency targets into buildings, as discussed in Chapter 6 and specifically Figure 6.2. Most European building occupiers will have come into contact with the EU Energy Labelling Directive, whereby appliances such as lightbulbs, computers, washing machines, dishwashers, kettles and toasters are rated in accordance with energy consumption. All these appliances can contribute to a reduction in energy demands, and may themselves promote more energy efficiency.

When energy consumption has been reduced significantly, it may then become more viable for domestic renewable energy systems (such as photovoltaic cells, solar thermal and wind turbines) to meet the full energy demand for the building.

In Sheffield, 2 St Pauls Place is the winner of the 2013 Chartered Institution of Building Services Engineers (CIBSE) Building Operation Award. The building's owner wanted to improve the building's environment and energy performance in order to reduce operational costs. A consultant was appointed to investigate how to achieve the owner's objectives. As a case study, this demonstrates how post-occupancy behaviour can influence design decisions (Box 4.5).

4.4.   Summary: collective impact on sustainable buildings

To make real changes to buildings in use, the occupants' needs should be identified at the beginning of the design process. Designing for adaptive occupant use is about considering the context on both the macro and the micro scale, and considering how and where people live, work and play. It requires the facilitation of an interaction of people with their environment, as well as their education, to ensure low carbon outcomes.

Maximising positive behavioural change for society can be influenced by creating an urban environment to provide people with a choice of sustainable transport, water, waste and power infrastructure. Then there is the choice of adapting to how people choose to lead their lives, by providing buildings that allow energy-responsible living, work and play. The combination will lead to lifestyles where the 'in-use' component of a building's life cycle can be optimised. The result of designing for adaptive occupant behaviour could offer people information as to exactly how they contribute to their community

Box 4.5 St Pauls Place, Sheffield, UK – how post-occupancy behaviour can influence design decisions (see Figure 4.4)

The objectives

The owner wanted to improve the environment and energy performance of the building in order to reduce operational costs. To this end a consultant was appointed to investigate how to achieve the building owner's objectives.

The solution

To retrofit the existing building services by, among other measures:

- improving lighting
- installing CO_2 monitors and controls on the fresh air ventilation systems
- adding thermal wheel heat reclaim devices to air-handling units to minimise heating losses
- providing IT server virtualisation to stop the use of individual desktop computers
- investing in photovoltaic cells and solar thermal systems for hot water.

Most importantly, the building users were fully informed, regular meetings were held and user surveys were carried out.

The building was the winner of the 2013 CIBSE Building Operation Award.

However, there is a lesson to be learned: if the building tenants and owner had had greater engagement in the original design process, then it is possible that their desired low-energy operation objectives could have been met from the outset of the building's design, saving time and costs.

Figure 4.4 St Pauls Place viewed from the outside. (Courtesy of Mott MacDonald)

economically, socially and environmentally, and thus help to provide a significant contribution to achieving an 80% reduction in global carbon emissions by 2050.

In conclusion, designing with 'in-use and end-of-life' in mind is a design philosophy that can be applied globally to create buildings with responsible outcomes that inform their occupants about how to follow a sustainable lifestyle.

Designing for adaptive occupant behaviour is currently being researched by many international groups, so some starting-point references are given below:

- The Why Factory: http://www.thewhyfactory.com
- The Future Cities project: http://futurecities.org.uk
- Technology Strategy Board (now Innovate UK) (2010) *Design for Future Climate. Opportunities for Adaptation in the Built Environment*: https://www.gov.uk/government/publications?departments%5B%5D=innovate-uk&page=2
- Rocky Mountain Institute: www.rmi.org
- The Smart Urbanist: http://www.smarturbanism.org.uk
- The Usable Buildings Trust: http://www.usablebuildings.co.uk

---

What can engineers do

- Consider the operational impacts of design during planning, and carry out investigative studies on energy, waste and transport.
- Engineers should get involved at the planning stage so that 'in life to end use' aspects of engineering can be taken into account.
- When designing integrated building systems keep in mind social needs and the possibility that the building occupiers may change within short time periods.
- Link your building to the urban context to facilitate operational use from the occupiers' perspective. How can you improve community connectivity?
- Consider flexibility, demountability or extra capacity in structural solutions.
- Can the design approach be long life; loose fit? Consider the life-cycle value of standardised and bespoke solutions, and apply to mechanical and engineering (M&E) systems and structural solutions. Can the M&E systems be easily stripped out when there is a change in building use? Can the structural solution manage changes in loading patterns?
- Review existing building stock. What can be done to reuse buildings and how can they be retrofitted? (Also adopt this approach to design for future use in new builds.)

REFERENCES

Ainger C and Fenner RA (2014) *Sustainable Infrastructure: Principles into Practice*. ICE Publishing, London, UK.

Arup and C40 Cities Climate Leadership Group (2011) *Climate Action in Megacities: C40 Cities Baseline and Opportunities*. Available at http://www.arup.com/Publications/Climate_Action_in_Megacities.aspx (accessed 12/11/2014).

C40 Cities Climate Leadership Group (2013a) About C40. Available at http://www.c40cities.org/about (accessed 12/11/2014).

C40 Cities Climate Leadership Group (2013b) *C40 Cities: Eight Solutions to Address Climate Change*. Available at http://www.c40cities.org/eight-solutions-to-address-climate-change/scenes-fallback.html (accessed 12/11/2014).

Duggan J (2013) China's mega water diversion project begins testing. Available at http://www.theguardian.com/environment/chinas-choice/2013/jun/05/chinas-water-diversion-project-south-north (accessed 12/11/2014).

Government of Japan (1997) Law for the Promotion of Sorted Collection and Recycling of Containers and Packaging (Container and Packaging Recycling Law). Ministry of Environment, Government of Japan. Available at http://www.env.go.jp/en/laws/recycle/07.pdf (accessed 12/11/2014).

Grow NYC (2014) *Recycling Facts*. Available at http://www.grownyc.org/recycling/facts (accessed 12/11/2014).

King's Cross Skip Garden (2014) Available at http://www.kingscross.co.uk/skip-garden (accessed 12/11/2014).

Kraljevic A (2011) *Big Cities, Big Water, Big Challenges: Water in an Urbanizing World.* WWF, Berlin, Germany.

Lewis M (2007) Megacities of the future. *Forbes*. Available at http://www.forbes.com/2007/06/11/megacities-population-urbanization-biz-cx_21cities_ml_0611megacities.html (accessed 12/11/14).

NCE (*New Civil Engineer*) (2012) How many holes to heat the Albert Hall. Urban scale sustainable heating and cooling. In *Infrastructure in 2012: A Special Report on Prospects for the Year Ahead*, pp. 28–31. Available at http://www.nce.co.uk/Journals/2011/12/16/d/d/x/Infrastructure-2012.pdf (accessed 12/11/2014).

UBT (Useable Buildings Trust) and BSRIA (2009) *The Soft Landings Framework – For Better Briefing, Design and Handover and Building Performance in Use*. BS4/2009. Available at http://www.usablebuildings.co.uk/Pages/UBPublications/UBPubsSoftLandings.html (accessed 12/11/2014).

UN Habitat (2010) *Solid Waste Management in World's Cities: Water and Sanitation in the World's Cities 2010*. Available at http://www.unhabitat.org/pmss/listItemDetails.aspx?publicationID=2918 (accessed 12/11/2014), p. xix.

United Nations (2012) *World Population Prospects: The 2012 Revision*. Department of Economic and Social Affairs, Population Division, Population Estimates and Projections Section. Available at http://esa.un.org/unpd/wpp/Excel-Data/population.htm (accessed 12/11/2014).

PWMI (Plastic Waste Management Institute, Japan) (2012) *Plastic Products, Plastic Waste and Resource Recovery*. PWMI Newsletter No. 41. Available at http://www.pwmi.or.jp/ei/siryo/ei/ei_pdf/ei41.pdf (accessed 12/11/14).

UKGBC (UK Green Building Council) (2013) *Key statistics*. Available at http://www.ukgbc.org/content/key-statistics-0 (accessed 12/11/2014).

Sustainable Infrastructure: Sustainable Buildings
ISBN 978-0-7277-5806-4

ICE Publishing: All rights reserved
http://dx.doi.org/10.1680/sisb.58064.083

ice
Institution of Civil Engineers

publishing

Chapter 5
Managing the process

Tristram Hope

> Here we must run as fast as we can, just to stay in place. And if you wish to go anywhere, you must run twice as fast as that.
>
> *Alice in Wonderland*, Lewis Carroll

5.1. Introduction

Anyone who has been involved in a construction project from its beginning to its end will be aware that building things, whether on the scale of a domestic self-build or a major international development, involves a complicated and often bewildering set of considerations, judgements, decisions and actions. Several specialist disciplines will typically be involved in helping clients to meet and overcome the many challenges along the route to achieving the optimum outcome.

It is therefore important to consider how the whole process can best be managed in order to achieve the most sustainable end result possible. It is also important to recognise that, apart from fundamental technical knowledge, the determination to drive rapidly-developing change, the ability to take a leading role with confidence, and the skills needed to communicate effectively using a variety of media are all essential tools for today's engineer.

The increasing use of internet-based file-transfer protocol (ftp) sites poses a potential risk, in that they can become a 'reactive' communication method if project team members fall into the trap of assuming that once information is uploaded onto the ftp site the rest of the team will automatically be fully aware of its presence and its implications. For this method of working to be successful, it has to be supported by 'proactive' communication, where the team member uploading the information makes an explicit effort to contact the rest of the team to ensure that they have received the information, accessed it, understand its relevance and know whether any following action is needed on their part. All too often, time and other commercial pressures tend to discourage this.

This chapter explores the way in which the many activities involved in the construction of a building can be broken down into stages, so that the necessary information required to proceed can be sourced at the appropriate time to achieve the desired end result. Timing is critical; the strategic philosophy of sustainability needs to be embedded in the project at the outset. Failure to do so will result in repeated opportunities to influence the design in a beneficial way financially, socially and environmentally being lost.

This chapter also discusses the work done recently by organisations in the UK to provide harmonised managerial guidance that takes due account of sustainability, and that will help project teams to progress through the construction process with a minimum amount of wasted effort and missed opportunities.

5.2. Workstages
5.2.1 What are workstages and why are they needed?

Workstages are defined, progressive steps in the construction process that describe the technical and managerial activities which need to be undertaken at different times to ensure that there is clarity between the project participants as to what needs to be done for best effect, and when. The financial management of the project can also be married into the workstages, with payments being released to the project team as specific sections of work are completed. A clear understanding of workstages is, therefore, essential within the project team.

5.2.2 Comparison of workstages

Table 5.1 provides an approximate comparison of the workstages recognised in the UK, France, Russia and the USA – these being established systems of working currently in widespread use globally. It can be seen that, while there are differences in the detail, there are also strong similarities, in that there are typically three main developmental stages, namely the initial inception and definition of the project, technical design work, and construction.

The importance of ensuring the continued involvement of the project team into the operational stage of the project, in the interests of gaining useful feedback and adjusting operational activities accordingly, has become widely recognised only relatively recently. The introduction by RIBA of a new Stage 7 (In Use) in their recent *RIBA Plan of Work 2013* (RIBA, 2013) acknowledges the need for activities such as carrying out post-occupancy evaluation, reviewing project performance, identifying and recording project outcomes, and possibly carrying out further research and development activity, using the building as a test-bed.

In the interests of greater carbon efficiency, it is no longer considered acceptable to design and construct buildings using a 'deemed to satisfy' approach. It is now expected that buildings will be designed to meet specific energy-performance requirements, and that these requirements can be demonstrably proven to have been met. The gathering of data during the operational stage of the project is essential in enabling the project team to evaluate particular features of the building, to see whether they actually do perform as anticipated, and therefore whether in reality they have a beneficial effect as far as the overall sustainability of the project is concerned.

5.3. Form of contract
5.3.1 The importance of collaboration

Efficient interaction and communication between team members has always been important to the successful completion of construction projects. However, the increased number of diverse participants who now need to be involved to ensure sustainable project

Table 5.1 International systems – approximate comparison of workstages

	0	1	2	3	4		5			6	7
UK (RIBA, 2013)	Strategic definition	Preparation and brief	Concept design	Developed design	Technical design		Construction			Handover and close-out	In use
France	ESQ	APS		APD	PRO	ACT	EXE	DET	OPC	AOR	
	Esquisse	Avant-projet sommaire		Avant-projet définitif	Projet	Assist. – contrats de travaux	Exécution	Direction de l'exécution	Ord., co-ord. et pilotage du chantier	Assist. lors des opérations de réception	
Russia	Konseptsiya	P		E	R		K				
		Proekt		Expertisa	Rabotchi Proekt		Konstruktsiya				
USA	Briefing	SD		DD			Construction			Warranty period	
		Schematic design		Detail design							

Box 5.1 British Airports Authority and the T5 Agreement

The construction of Terminal 5 at Heathrow Airport benefited greatly from the adoption of a holistic partnering approach. Appreciating the highly complex nature of the project and the many significant technical difficulties and risks involved with it, the client (British Airport Authority) realised the importance of assembling a highly focused and determined construction team that could perform to the best of its ability without the obstacles and distractions typically brought about by conventional procurement methods.

The Terminal 5 project varied from the 'business-as-usual' approach in that risk was managed by the client rather than being transferred to the contractor. Properly incurred costs were reimbursed, instead of prices being fixed in advance, and profit levels for discrete packages of the works were negotiated equitably instead of being held at risk. Performance of the project team was success-driven, instead of penalty-driven, and the client, consultants and contractors were co-located as far as possible in order to ensure an optimum level of integration and to discourage 'silo' working. The expectation was for exceptional performance from all participants, instead of merely best practice being sufficient.

A bespoke partnering contract known as the 'T5 Agreement' was written to enable this. It assumed and allowed for the following:

- involvement of all suppliers from the outset
- co-located teams
- open-book accountancy arrangements with cost-reimbursable contracts
- incentive schemes with time and cost targets
- project insurance policies, with suppliers paying shares of the excess
- ring-fenced profit.

The T5 Agreement worked well, and the resulting culture of honesty, trust and mutual confidence ensured the successful delivery of the project from economic, social and environmental points of view.

In addition to the contractual aspects, the issue of construction waste was also addressed very effectively, partly by adopting an off-site manufacturing approach where possible, in order to minimise initial wastage, but also by providing simple yet effective means of collecting, segregating and recycling of site-generated construction waste. As a result, overall recycling levels of up to 97% were claimed to have been achieved for the project.

---

outcomes means that collaboration is crucial. It is therefore important to instil a sense of shared benefit and ownership among all project participants, and to make every effort to encourage engagement and involvement of all parties. This was achieved to great effect in the project to construct the new Terminal 5 building and ancillary works at Heathrow Airport (Box 5.1).

UK Government Advisor Sir Michael Latham raised the issue of 'fragmentation' within the UK construction industry in his report *Constructing the Team* (Latham, 1994). One of the proposals he made in his report was that the traditional client/contractor approach to project procurement, which readily led to adversarial relationships within the project team, should be replaced by a 'partnering' approach:

> Partnering includes the concepts of teamwork between supplier and client, and of total continuous improvement. It requires openness between the parties, ready acceptance of new ideas, trust, and perceived mutual benefit…

5.3.2 The development of the NEC suite of contracts

A crucial choice that can significantly assist the development of a collaborative attitude relates to the type of contract to be used. Originally launched in 1993 as the 'New Engineering Contract', the NEC suite of contracts was praised by Latham for its collaborative and integrated working approach to procurement. It was strongly recommended by him as the best example of a modern construction contract that embraced partnering.

It is now in its third edition, known as NEC3 (2013), and covers a wide range of possible construction scenarios, as demonstrated in Figure 5.1.

5.4. Range of activities
5.4.1 The CIC scope of services

To create a partnering culture such as that advocated by Latham, it is important that all project participants have a clear understanding of what is expected of them and what services they are to perform. Several systems are available that describe the scope of work

Figure 5.1 The NEC3 suite of contracts and its range of application. (Courtesy of NEC)

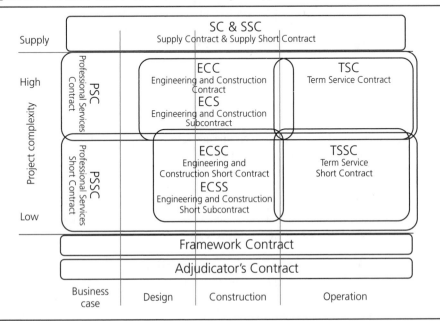

to be undertaken by different disciplines but these are not explicitly harmonised one with another. Scopes of service that are oriented towards particular design disciplines tend to lead to 'silo' thinking, as well as leaving the possibility of responsibility overlaps or, worse still, the exclusion of important activities that need to be performed.

This problem was recognised several years ago within the UK construction industry by the Construction Industry Council (CIC), with the result that a holistic scope of construction services document was drawn up, known as the *CIC Scope of Services* (CIC, 2007). It comprises integrated and detailed lists of services that may be required on construction projects, from inception to post-practical completion. From these lists, schedules of services can be drawn up for the appointment of consultants, specialists and contractors, thereby allowing the complete, transparent and unambiguous definition of the activities to be undertaken.

The document is intended for use on major building projects designed by multi-disciplinary teams, irrespective of the procurement route to be used. The client may be a developer, the end-user, or a design-and-construct contractor. The services may be undertaken by any or all of the following:

- consultants (engaged under the CIC Conditions section of the *CIC Consultants' Contract* (CIC, 2011), or on any other basis)
- specialists providing design input
- contractors (either under a traditional or a design-and-construct procurement route).

The document refers to the 'definition' of a project, rather than to its 'design', because far more than design is involved and not all of those participating in the process undertake design activities. Thus the process includes management, coordination, costing, programming, health and safety, and so on.

Thus:

Definition + Construction = Project.

A series of tables details the tasks to be undertaken by all members of the project team as part of the 'definition process', and the document sets clearly-defined stages for the refinement of the project definition.

So a partnering form of contract, such as one of the NEC suite, supported by clear activity definitions based on *The CIC Scope of Services*, provides a sound contractual framework on which to base sustainable projects.

5.5.   Activities in support of sustainability

All participants have a part to play in order for sustainable outcomes to be achieved, so it is important to provide suitable opportunities for all parties to be able to voice their views, however radical they may seem, effectively and without embarrassment. It is

Box 5.2 The RIBA 2011 *Green Overlay* and *Plan of Work 2013*

In 2011, RIBA published a *Green Overlay to the RIBA Plan of Work* (RIBA, 2011), which provided *aide memoire*-style guidance to help project participants ensure that critical sustainability issues were raised and addressed at appropriate times during the project. This guidance was subsequently embodied in the *RIBA Plan of Work 2013* (RIBA, 2013), which includes references to 'sustainability checkpoints' to be reviewed at each of seven key project stages.

therefore important to create an atmosphere of trust, respect and cooperation within the project team.

This can be achieved by implementing a series of meetings at different stages through the project, preferably starting as early as possible. This was proposed by the RIBA in its *Green Overlay to the RIBA Plan of Work* (RIBA, 2011), as outlined in Box 5.2.

Table 5.2 provides a simplified summary of the more major issues to be taken into account at each of the seven stages proposed by the *RIBA Plan of Work 2013*. While the guidance was initially written with the architect in mind, it is equally useful for all project participants; indeed, it is important that all members of the project team are aware of it, and understand its implications.

The practical way in which this guidance can be implemented is through holding a series of meetings (shown in bold in the 'Activity' column), typically as an adjunct to regular project progress meetings, during which those specific actions that relate to sustainability can be highlighted.

Table 5.2 Summary of sustainability activities, based on the *RIBA Plan of Work 2013* (RIBA, 2013)

Workstage	Activity	Technical/contractual/managerial
0: Strategic definition	**Initial sustainability meeting** Assess client's level of awareness of the economic, social and environmental aspects of sustainability. Discuss fundamental, operational and personal principles (see Chapter 2).	Discuss the project purpose and target outcomes. Identify: key drivers, such as the rate of return on investment and the Capex/Opex balance; possible corporate social responsibility issues; and possible environmental impact issues. Encourage the adoption of a partnering form of contract and an appropriate procurement route. Check the progress of legal due diligence work.

Table 5.2 Continued

Workstage	Activity	Technical/contractual/managerial
1: Preparation and brief	**Sustainability workshop** Develop personal reference frameworks. Using the data gathered, produce the sustainability action plan. State aims and aspirations for improvement over the existing situation.	Establish the core team with the necessary appropriate technical skills. Agree primary communication contacts. Encourage multidisciplinary interaction. Identify and organise any necessary investigative work and local area surveys. Define any environmental impact control measures as appropriate. Agree on the assessment methodology to be used. Allow sufficient time in the project programme for repetitive cycles of design, review and adjustment. Ensure all parties sign up to the sustainability action plan.
2: Concept design	**Sustainability Assessment 1** Benchmark the existing situation. **Sustainability strategies meeting** Ensure that the implications of the sustainability strategies to be implemented are understood by the project team. Site waste management plan to be discussed and developed.	Assess the level of sustainability of current operations. Develop the project, ensuring that the layout respects and facilitates sustainable design strategies. Ensure continuous collaborative discussion across all disciplines and within the project team as a whole. Compare possible solutions that suit key drivers, and check their appropriateness. Check anticipated Capex/Opex figures.
3: Developed design	Develop specific features to enable sustainable design strategies. **Sustainability Assessment 2** Assess the sustainability of the proposed project. Compare it with the existing situation and amend the project if necessary.	Check that the sustainable design strategies satisfy the previously agreed aims and aspirations.
4: Technical design	Produce the building user guide, including adaptation and deconstruction strategy documents. Develop strategies to influence building user behaviour.	Develop sustainability management and monitoring techniques.

Table 5.2 Continued

Workstage	Activity	Technical/contractual/managerial
5: Construction	Monitor site waste management. Implement site sustainability procedures.	Monitor and test sustainable design features to ensure suitable levels of performance.
6: Handover and close-out	**Sustainability assessment 3** Carry out the as-built project sustainability assessment and compare it with the existing situation assessment and proposed project assessment. Record outcomes.	Hold a project close-out review and take note of feedback gained for future reference, for dissemination within the project team, and more widely if possible. Ensure that all relevant information gathered throughout the project, especially copies of all investigative reports, operation and maintenance manuals, building information modelling information and the building user guide, is recorded and copied to the principal project team members and building manager(s)/operator(s).
7: In use	**Sustainability assessment 4** After the first year carry out an operational sustainability assessment, and disseminate the outcomes to project team members. **Sustainability monitoring** Carry out periodic operational assessments, and review these against the original aims and aspirations.	Use information gathered in the operational sustainability assessment to identify any necessary changes to building operational procedures or building occupant behaviour patterns. Implement strategies to change operational methods or behaviour, possibly including building user induction sessions.

What can engineers do?

Engineers are well placed to organise and manage the process of defining and constructing projects, by virtue of their typical skill set, which includes the ability to sort and organise data, to evaluate new ways of doing things and adopt or reject them as appropriate, to devise practical solutions based on theoretical assumptions, and to communicate their ideas effectively. In order to maximise the chances of achieving sustainable outcomes, engineers can:

■ As well as developing fundamental technical knowledge, emphasise communication and leadership skills, in order to be able to take a leading role in driving sustainability issues throughout the project process.

- Develop an awareness of the differences and similarities between the various construction workstages used in different jurisdictions, in order to be able to take appropriate action in support of sustainability at the optimum moment in the project.
- Encourage the project team to adopt a collaborative partnering type of contract, in order to facilitate a project culture of trust, mutual confidence and open-mindedness.
- Ensure that all aspects of the project are defined unambiguously and in sufficient detail, and that activities are allocated to those best suited to undertake them.
- Organise and implement a framework of meetings specifically to ensure that key sustainability activities are undertaken at the appropriate stages in the project.

REFERENCES

CIC (Construction Industry Council) (2007) *The CIC Scope of Services*. RIBA Publishing, London, UK.

CIC (2011) *CIC Consultants' Contract Conditions*, 2nd edn. RIBA Publishing, London, UK.

Latham M (1994) *Constructing the Team*. Department of the Environment, London, UK.

NEC3 (2013) *NEC3 Suite of Contracts*. Thomas Telford, London, UK.

RIBA (2011) *Green Overlay to the RIBA Outline Plan of Work*. RIBA Publishing, London, UK.

RIBA (2013) *RIBA Plan of Work 2013*. RIBA Publishing London, UK.

Sustainable Infrastructure: Sustainable Buildings
ISBN 978-0-7277-5806-4

ICE Publishing: All rights reserved
http://dx.doi.org/10.1680/sisb.58064.093

Chapter 6
Assessment methodologies, targets and reporting requirements

Alan Yates

> Friend to Groucho Marx: *Life is difficult.*
> Groucho Marx to friend: *Compared to what?*

6.1.   Introduction

Sustainable building design covers a wide range of interrelated issues, and the introductory volume in this series explores these in more detail (Ainger and Fenner, 2014). This complexity makes evaluation of sustainability impacts difficult and time-consuming to consider in full for an individual building development. Few buildings are designed and constructed by a single person or organisation, and, as a result, projects typically involve a high level of communication between the design and construction team members in order to deliver the final building. This requires careful and informed management at many levels and the need to measure and communicate in a way that will be easily understood and actioned by all parties.

The design process is often one of risk management and compromise between a range of potentially conflicting issues and complex interactions. Simplified approaches are needed for the measurement, reporting and communication of information on building performance and the impacts of design decisions. This information is needed in a timely and accessible fashion so that collaborative decisions can be made at an appropriate stage in the design and procurement of a building. Credible and comparable metrics and methods are critical, so that options can be properly considered.

There are many approaches and tools available to meet the needs of different stakeholders for evaluating building performance, and the situation is changing all the time. As a consequence, this chapter cannot provide a comprehensive guide to each of them but instead outlines the key principles of selecting and using methods appropriately to meet the needs of the project and its stakeholders.

6.2.   Why measure?

Decisions are often taken based on a set of assumptions that later prove to be inaccurate, or where subjective perceptions are used to make choices that fail to achieve the desired outcomes. This leaves any decision open to questioning, and ultimately to criticism, and will often create significant risks of failure and, potentially, liability.

Failure to fully understand potential outcomes and impacts at an appropriate stage in a project can lead to changes in direction later that may have a major impact on the ability to meet objectives, or on the cost involved in doing so. This is true of any project but is particularly important for those that seek to push the boundaries. The earlier an unforeseen impact is identified, the easier and less costly it will be to overcome.

6.3. Setting targets

Performance targets are frequently set for buildings by clients, funders, regulators and local authorities early in the briefing stage. Designers and constructors may also set internal targets in line with their own policies and procedures, which are often used to demonstrate capabilities when bidding for future work.

Establishing a holistic set of clear targets early on in a project provides clarity for all members of the design and construction team, and allows easy evaluation of success later on. While there may be a place for some discrete performance targets for issues such as energy use, setting too many single-issue targets can stifle creativity and innovation in a design process, risking the creation of conflicts that cannot be resolved. Costs may also be increased by limiting the ability to find cost-effective options and balance priorities as a design develops. It is also possible to over-measure by setting targets without a clear benefit. Both of these should be avoided. Measurement is only beneficial when it serves a purpose, and more holistic evaluation tools overcome this problem by identifying key issues and comparing and balancing the importance of these issues to the environment, occupiers and property value.

Targets can be set and performance monitored based on previous experience or, alternatively, on industry benchmarks. However, this can lead to the application of overly-generic solutions, so care must be taken to ensure that the benchmarks are appropriate, as the context of each building is likely to differ significantly. Another option is to use a whole-building evaluation tool such as BREEAM or LEED to provide an appropriate degree of flexibility, allowing the design and construction team to accommodate the project's context. These methods are generally based on the principle of a balanced scorecard to allow a high degree of flexibility and focus on the measurement of *impact*, rather than being solutions based and prescriptive in their requirements (Box 6.1). They are based on established industry benchmarks in their respective countries or territories, and so care must be taken to ensure that these are appropriate to the location of the building. BREEAM allows for standards and benchmarks to be adapted to meet local requirements and regulations, albeit within a common framework and scope, to ensure a degree of comparability between assessments across countries.

Box 6.1 A balanced scorecard

A balanced scorecard is an evaluation and management system used to align design decisions across a diverse range of project objectives, issues or impacts. It provides a framework to guide a project team through complex decisions by establishing the relative importance of diverse issues and criteria, usually based on consensus.

Setting single-issue targets that address one particular impact or aspect of a design can be helpful in providing the design team with a clear objective. However, they encourage isolated consideration, which can, and often does, result in unforeseen consequences that are difficult or expensive to overcome. Setting too many discrete single-issue targets runs the risk of oversimplification of a complex problem, potentially creating conflicts, and may also result in higher costs and unforeseen consequences. A more flexible and robust approach would be to set an overall target based on a holistic building evaluation method and a small number of specific targets for priority issues, such as energy consumption or thermal comfort (Box 6.2).

Box 6.2 Setting sustainability targets for the London Olympics 2012 (Commission for a Sustainable London, 2012)

The report *London 2012 – From Vision to Reality*, published in November 2012 (Commission for a Sustainable London, 2012), analysed the successes of the strategy of the London Organising Committee for the Olympic Games (LOCOG) in setting appropriate targets, driving these through design and procurement, and in monitoring and steering decisions made throughout the period from concept and bidding to final delivery of the Games in the summer of 2012. This strategy focused on both the 'games-time' sustainability objectives, targets and aspirations, and the 'legacy' sustainability through reuse and adaptation of venues (Figure 6.1). The report concluded that London 2012 met the majority of its targets and succeeded in its objective to deliver the most sustainable Olympic Games ever.

Figure 6.1 The Olympic park, London, UK. (Courtesy of the ODA)

The report concludes that the:

> many partners responsible for staging the Games were provided with the best possible platform by the Olympic Development Agency (ODA). All of the venues and the Olympic Village were successfully constructed to the highest sustainability standards with unprecedented levels of energy and water efficiency; well designed and constructed using sustainable materials. The infrastructure underpinned this commitment. The use of combined cooling, heat and power and black water recycling ensured that energy and water were not only conserved, they were supplied from more sustainable sources. The final piece of scene setting was the visionary design and delivery of the natural environment by the ODA. This not only provides Europe's biggest new urban green space for 150 years but also provided a stunning natural backdrop to the world's premier event.

The ODA worked with BRE to adapt BREEAM to drive sustainability in the design of the venues. All permanent venues were designed to achieve a rating of Excellent. In addition, the athletes' village was designed to meet the requirements of the Code for Sustainable Homes level 4. The ODA also used CEEQUAL to drive good environmental management practices across the broader infrastructure design.

6.4. The art of measurement

6.4.1 Who needs to measure?

Different stakeholders have differing interests in terms of the measurement and presentation of a project's impacts.

- *Property funders* will want to ensure that their investments are being made in a way that protects the value of their assets.
- *Developers and clients* will want to be able to communicate their priorities and evaluate and demonstrate the success of their design and construction team in achieving these.
- *Designers and specifiers* will want to manage risk by effectively and efficiently informing their decisions, justifying proposals and demonstrating compliance and successes.
- *Constructors* will want to understand underlying requirements and confidently manage their choices on the basis of these, as well as demonstrating compliance, whether in respect of regulatory or contractual requirements.
- *Occupiers* will want to have confidence that the property they occupy, buy or rent meets functional and economic needs, while complying with policies of corporate social responsibility and environmental management.
- *Regulators* need to use clear and transparent methods to ensure that statutory requirements are being met in a fair and equitable way.
- *Manufacturers and suppliers* throughout the supply chain need the means to demonstrate in a comparable, credible and independent manner that their products meet the project requirements.

All of the groups above will measure for different reasons, but the use of a common evaluation method simplifies the role of the design team in servicing these demands, and ensures that comparisons and contradictions can be explored adequately by a range of stakeholders.

6.4.2   What to measure?

In order to have a credible understanding of the environmental or sustainability impacts of a building, the following need to be considered:

- Procurement impacts, including resource efficiency, construction energy/water, disturbance, construction waste and pollution.
- Building-related operational impacts, including operational energy and water consumption, waste and recycling, internal environmental parameters such as lighting, ventilation and comfort.
- Broader operational impacts relating to transport and a building's impact on local climate, community, employment and economic activity, biodiversity/ecology and pollution levels.
- Quality, adaptability and credibility of the design and procurement processes adopted.

The detail behind these generic categories will depend on a project's nature, location and use. Using a credible tool to measure impact removes the need to produce a project-specific set of issues to measure against, thus making the task manageable, robust and easily understood by others. This approach should not preclude the identification of project-specific issues or priorities where these are not covered by the generic system, as long as this is done in moderation and can be justified.

6.4.3   When to measure?

To be effective in ensuring an optimal outcome, an appropriate level of measurement and benchmarking of performance should be carried out throughout the design, construction and operational stages of a building's life. Most designers and constructors attempt to do this internally to some degree but approaches will differ between organisations and so will be less useful in overall project management terms as outcomes will not be directly comparable.

What is measured and who needs to be involved will vary but it is important to ensure that the decisions made are being based on credible comparisons, so that the ability to meet project objectives is not inhibited by restricting flexibility unnecessarily. BRE, CIRIA and others have published a range of guidance on the appropriate stages for carrying out measurement, comparison and feedback within the design process, including the optimal timeline and its potential impact on the outcomes under a formal certification scheme such as BREEAM. Generally, the earlier the information gathering process can begin the better.

6.4.4   What to evaluate

Many design/evaluation tools and guides available for the construction sector consider individual aspects of a building's impact or performance in isolation. At the design stage this is helpful in informing the comparison of different options, but is in itself inadequate in making a final decision on which option to adopt. It is necessary to understand how such choices impact on other issues. To do this, and to ensure that unforeseen consequences are avoided, a more holistic approach is needed.

Consideration of a single issue in this way not only makes assumptions about use patterns and assessment boundaries internally within the tools to simplify the calculations, but also takes no account of the impact of the issue on other factors. For instance, the standard assessment procedure (SAP) and standard building assessment method (SBEM) calculation methodologies consider the thermal heat transfer through an external wall or roof but do not consider the environmental or cost implications of a particular insulation material used to achieve this. A high-performing building in energy terms may fail to perform well in terms of thermal comfort or ventilation, unless designers make sure that they properly understand and evaluate these resultant impacts to ensure that a building performs well holistically. These interactions often have a major impact on the success of a building in operation, and therefore on the reputations of those involved in the project. Using a holistic design or evaluation tool helps to avoid accidentally overlooking key performance areas at a critical time in the design process.

6.4.5 Actual versus predicted performance

At the design stage, any evaluation of performance has to be based on a set of assumptions about how the building will be occupied, managed and maintained. User behaviour has a large impact on the actual performance of a building. This is especially true for a building that operates in a low-tech or simple passive manner. Design prediction tools are often criticised when actual performance in use fails to meet the predictions (Box 6.3). This is to be expected to some degree but the accuracy of the predicted performance will depend both on the robustness of the tool used and on the quality and accuracy of the data that has been entered. Some tools are based on historic performance data in use, whereas others are based on theoretical performance modelling. Both approaches have strengths and weaknesses, and should be seen as useful tools within their limitations.

Box 6.3 Design tools – actual and predicted performance in use

Building regulations, BREEAM and other compliance or rating tools often use standard methods for the calculation and benchmarking of the potential energy consumption of a building design. In the UK, SAP is used for domestic buildings and SBEM is used for non-domestic situations to provide a common national calculation method for energy performance and resultant carbon emissions. These tools should not be used to give an accurate prediction of actual energy consumption, and as such should not be used as detailed design tools. Energy modelling of complex buildings within the design process is usually carried out using dynamic modelling software, which is capable of considering a wider range of factors and allows a high degree of sensitivity analysis to be carried out. This is important in determining the robustness of design decisions and the ability of a building design to adapt to differing demand scenarios. They do not, however, allow for broad comparison in terms of compliance due to the flexibility that these tools permit.

6.5. The principles of credible building performance evaluation methods

The following principles underpin all credible evaluation tools and techniques.

6.5.1 Quantification

Wherever possible, evaluation should be carried out on the basis of robust quantification of the actual or predicted impacts. This allows direct comparison rather than subjective value judgement. For impacts such as energy consumption or the embodied impacts of materials this is a feasible task if established tools are used. Other impacts are less easily measured, and the evaluation may be more subjective. In such cases it is still important to use a broadly accepted method and metric to allow fair comparison and a robust outcome.

When comparing quantified impacts care must be taken to ensure that the quoted metrics are comparable. Failure to do this can lead to decisions being reached that fail to achieve the desired outcomes, as illustrated famously by NASA in 1999 – in that case resulting in an expensive and irrecoverable error (Box 6.4).

While an extreme example, the NASA case illustrates the impacts of a failure to check the quality of evaluation processes properly. Data are often provided in varying forms and obtained using differing metrics. This may not be immediately obvious to the uninitiated, but the consequences can be significant, difficult and expensive to rectify.

6.5.2 Independence and credibility

Given that evaluation is used for a wide range of purposes, it is important to consider the appropriate level of independence required. To demonstrate compliance with regulations or a third-party assessment method such as LEED or BREEAM, an independent

---

Box 6.4 NASA metrics error (Millar, 1999)

'Pasadena, California (Reuters, September 1999). Human error stemming from space engineers using two sets of measurements – one utilizing miles and the other kilometers – caused the loss of the Mars Climate Orbiter spacecraft last week, NASA said Thursday.

The teams, located at the National Aeronautics and Space Administration's Jet Propulsion Laboratory in Pasadena and at Lockheed Martin Astronautics in Colorado, complicated matters further by failing to realise the error, the agency said in a statement.

The $125 million orbiter, intended to serve as the first interplanetary weather satellite, is believed to have broken up when it hit the Martian atmosphere last week after an approach that was too near the surface.

"People sometimes make errors," said Edward Weiler, NASA's associate administrator for space science. "The problem here was not the error, it was the failure of NASA's systems engineering, and the checks and balances in our processes to detect the error. That's why we lost the spacecraft."'

---

evaluation is required to give confidence in the results. Similarly, an organisation will require a third-party review, where the outcomes are to be reported externally to demonstrate competence to the market or for use in corporate social responsibility reporting.

This is not the case when evaluation is being undertaken to inform design decisions, where a self-assessment approach may be more appropriate, as long as the user is able to understand the level of rigour required in inputting data and is able to interpret the outputs to avoid misunderstandings. It is easy for self-evaluation to mislead if this understanding is not present.

6.5.3   Evidence base

To be credible, an evaluation method should be based on sound scientific evidence and a clear understanding of the benefits of the decisions being taken. In reality this science base is incomplete in many areas, and it is important that data are collected to improve our understanding of buildings in use. A range of organisations, including consultants, national research bodies such as BRE, and non-governmental organisations (NGOs) such as the Carbon Trust and the Waste & Resources Action Programme (WRAP), seek to collect quantified and comparable data. BREEAM collects and reports data on a range of key performance indicators and uses the data to inform future development of the method. BRE is actively looking at how these data can be made more widely available to inform decisions and tool development in the future.

It must be accepted, however, that most methods make assumptions and set boundaries in both their criteria and scope. This means that actual performance will almost always differ from that predicted. This is especially true of issues affected by occupant behaviour, which cannot be accurately foreseen at the design stage.

It is widely accepted that the fuel-consumption figures quoted by car manufacturers are, at best, only indicative of performance, and actual fuel consumption may be significantly different as a result of driver behaviour and the operation of other systems in the car, which are isolated during bench testing. The market accepts this for cars but seems to have a problem recognising that the same applies to buildings. A frequently quoted issue relates to the significant variability in energy performance ratings required under the EU Energy Performance of Buildings Directive (European Commission, 2002). The energy performance indicated by the energy performance certificate (EPC), and based on design-stage calculations using SAP and SBEM in the UK, is usually significantly different from that indicated by the display energy certificate (DEC) at the operational stage, which is based on actual measured consumption.

The rigour and accuracy of evaluation methods is also compromised by the need to ensure that evaluation is practicable and achievable. This limits the complexity of any assessment method, and often means that established measurement tools are utilised in certification schemes, as these calculations are already being carried out for other reasons. This is especially true of the holistic building evaluation methods mentioned earlier, which seek to minimise the need for design teams and manufacturers to carry out duplicate calculations to others that they are already required to produce. This is not a problem from the point of view of comparability and benchmarking performance but does suggest that they are not a complete substitute for more complex modelling during the design process, where this is warranted. Another concern is that this simplification may limit the ability to take account of more innovative solutions, although these are often taken into account elsewhere in the method, where they provide credible benefits. This is the reason why BREEAM incorporates credits relating to innovation, and to the dissemination of the outcomes.

6.5.4 Comparability

Where evaluation is being carried out to inform design or specification choices, demonstrate performance relative to others, or to measure compliance, it is more important for the outcomes to be comparable than it is for them to be fully accurate. This creates the need for common standardised frameworks against which assessment and evaluation can be made, and an acceptance that the output information is relative and not absolute.

When making comparisons of performance it is important that the user understands the potential variation in methods and boundaries that may be influencing the results quoted. A good example is in the area of life-cycle assessment (LCA). LCA is a procedure that aims to measure the environmental impacts of a product over its life across a wide range of indicators.

There are a number of different established frameworks under which an LCA evaluation may have been undertaken. A full LCA would consider all impacts at all stages, from extraction of raw materials, through their manufacturing, transportation, construction, operation, repair and maintenance, and end-of-life impacts, including the recycling of materials. This is often described as a 'cradle-to-cradle' LCA. Given the growth in recycling of construction materials in many parts of the world, including the UK, a 'cradle-to-grave' LCA is usually more informative. This analysis stops short of the recycling of demolition materials at end of life, which often results in double counting of impacts and benefits, and so can be misleading if not used carefully.

In reality this is not particularly meaningful for most construction products or building components, as many of these factors fall outside the influence of the manufacturer or designer. Commodities such as oil are sold through markets that make it impossible to trace the resource back to the original point of extraction. Similarly, the transport burden to site will be dependent on the site location relative to the manufacturer, and for a complex material or component local choices may not be available in many cases.

In such instances the scope of the LCA will often be limited to those stages up to the point at which the product leaves the factory gate ('cradle-to-gate' LCA). Such an analysis will not include any impacts occurring after the product leaves the factory. Impacts associated with the transport of materials to a construction site, the processes and wastage on site, and the impacts arising at end of life are not considered. This poses a problem for the designer: a product with a long life/low maintenance burden may appear to have lower impacts than a less durable one with relatively lower cradle-to-gate impacts. The choice should be made in favour of the option that has the lowest overall impact across the life of

---

Box 6.5 LCA assessment standards

LCA framework standards:

- ISO 14040:2006: Environmental management – Life cycle assessment – Principles and framework (ISO, 2006a)
- ISO 14044:2006: Environmental management – Life cycle assessment – Requirements and guidelines (ISO, 2006b).

Greenhouse gas product LCAs can also comply with standards such as:

- PAS 2050:2011 Specification for the assessment of the life cycle greenhouse gas emissions of goods and services (BSI, 2011)
- *Protocol Life Cycle Accounting and Reporting Standard* (Greenhouse Gas Protocol, 2013).

---

the building, and so in-life impacts must be considered alongside those quoted by a manufacturer or trade body based on a cradle-to-gate LCA.

An added difficulty occurs because, while agreed frameworks exist for the definition of a full cradle-to-grave LCA, no such agreed approach is in place for cradle-to-gate assessments and, as such, a specifier may find that the data quoted for different products are not comparable, and could mislead if taken at face value (Box 6.5).

6.6. Types of evaluation tools

These will usually take the form of a standard or code of practice, and there are many examples. They fall into a number of discrete categories, as follows.

6.6.1 Whole-building evaluation methods

These provide a holistic overview of performance and usually assess performance against discrete criteria, which are then brought together into a single score or rating using a transparent or hidden set of weighting factors.

Widely used examples of such methods include:

- BREEAM (Building Research Establishment Environmental Assessment Methodology) (UK, other national and international versions relating to building and community levels) – adopts relevant elements of CEN framework standards for the environmental evaluation of buildings (Box 6.6 and Appendix 1)
- LEED (Leadership in Energy and Environmental Design) (US and international use based on the US version, relating to building and community levels)
- DGNB (Deutsche Gesellschaft für Nachhaltiges Bauen) (Germany, Austria and Denmark) – adopts relevant elements of CEN framework standards for the environmental evaluation of buildings (see Box 6.6)
- Greenstar (Australia)
- NABERS (National Australian Built Environment Rating Scheme) (Australia)
- Pearl Building Rating System (Estidama – Abu Dhabi Urban Planning Council) (Middle East)
- Others relating to specific countries and often based on the above.

Box 6.6 Sustainability of construction works – UK national standards

CEN has published a suite of framework standards relating to the sustainability assessment of buildings in 2011–2012 through the relevant national standards bodies. The UK versions are as follows:

■ BS EN 15643. Sustainability of construction works. Sustainability of buildings.
Part 1:2010. General framework.
Part 2:2011. Assessment of buildings. Framework for the assessment of environmental performance. Environmental impact assessment.
Part 3:2012. Assessment of buildings. Framework for the assessment of social performance.
Part 4:2012. Assessment of buildings. Framework for the assessment of economic performance.
■ BS EN 15804:2012. Sustainability of construction works. Environmental product declarations. Core rules for the product category of construction products.
■ BS EN 15978:2011. Sustainability of construction works. Assessment of environmental performance of buildings. Calculation method.

Note: These standards are voluntary but both BREEAM and DGNB are working to bring their methods in line with the relevant requirements of these standards. Both schemes meet the majority of the relevant requirements with regard to their respective scope. While DGNB follows more closely the presentational structure set out in the standards themselves, the structure and outputs of BREEAM are aimed at making the process as simple to follow as possible, following more closely typical design and construction practices.

6.6.2 Discrete impact evaluation methods

These are developed and published/operated by a wide range of organisations, to assist in the modelling and reporting of specific aspects of a design (e.g. energy and daylight modelling software). They vary in terms of their scientific basis and degree of independence, and are usually focused on providing accessible design and specification assistance, or to allow specific auditing of consumption in order to predict future operating costs or other impacts.

A wide range of tools is available, and the level of robustness and rigour varies enormously. Care should be taken to ensure that a credible tool is being used that is fit for the purpose for which it is intended.

6.6.3 International, regional and national standards

■ *ISO, European (or other regional) and British (or other national) standards* are published by recognised international or national standards bodies. There are three distinct types of standard published:
  – *Performance standards* set specific performance criteria for products and services, and are used as the basis for performance tests carried out by manufacturers and testing organisations. Their intention is to provide an

acceptable level of confidence in the performance levels that will be achieved. They may provide a single minimum acceptable level of performance, or a graded one to allow for product differentiation. They are often very specific in their scope, such as BS 10621:2007 + A2:2012, 'Thief-resistant dual-mode lock assembly' (BSI, 2007).

- *Framework standards* seek to ensure a degree of comparability between those standards that comply with them. They will not ensure direct comparability, and as such are of limited direct value to those involved in the design or procurement of a building or its constituent parts. An example would be BS EN ISO 8902:2009, 'Responsible sourcing sector certification schemes for construction products' (BSI, 2009). Box 6.6 lists a set of framework standards that set underpinning requirements used by BREEAM and DGNB in determining the scope and rigour of their whole-building certification schemes.
- *Standards covering process or testing regimes* ensure that data provided by companies are comparable.

■ *Standards operated by independent standards bodies, including national research bodies* (such as BRE, CIBSE, HVCA and others in the UK) may mirror the options set out above but are often formulated to provide accessible product and service differentiation through market-facing ratings or scoring mechanisms. For example, BES 6001, 'The framework standard for responsible sourcing' (BRE Global, 2014), supports BS 8902:2009 (referred to above) by providing a system that enables construction product manufacturers to ensure and prove that their products have been made with constituent materials that have been responsibly sourced.

■ *Company or sector level specification and measurement tools* are often used to define and differentiate specific services and commercial offerings in a competitive marketplace, or provide an accessible means of comparing a company's products. Examples would be the comparative reporting by a manufacturer of the performance characteristics of its products, such as Environmental Product Declarations (EPDs), or a construction company's approach to environmental management of its operations on a site.

■ *Codes of practice* define an appropriate level of service, and as such say nothing in relation to the quantifications of outcomes themselves, although they can be invaluable in defining the levels of service required or provided.

■ *Guidance* is provided to enhance understanding and ease addressing of issues, but in itself is of limited value in reporting performance levels or communicating performance between design team members.

6.6.4 Regulatory compliance tools

A number of calculation/modelling tools are used in regulatory compliance. Examples in the UK would be the energy modelling tools SBEM and SAP mentioned previously. Such tools are often based on simplified models, which by their nature do not consider all aspects of a design, and as such cannot be considered accurate in predicting overall consumption. This is not a problem when they are used for their intended regulatory purpose, as long as they give a broadly representative answer. This limitation, however, means that such tools should not generally be used as design tools. As they usually compare design performance against a notional baseline based on typical performance, such tools may not fully recognise the benefits of design proposals where the underlying performance of the building is already above that of the typical benchmark. Certification schemes such as BREEAM will often make use of these compliance tools, for two reasons. Firstly, all stakeholders are required to use them to meet regulatory requirements, and so there is no additional time burden associated with demonstrating compliance with the higher BREEAM standards. Secondly, these methods are used to make comparisons and to benchmark performance, rather than to provide a design tool that can be used for option appraisal. In this respect they are similar in their needs to what is required to demonstrate regulatory compliance. It is important that outputs are comparable between projects so that they can be properly benchmarked against predicted consumption targets.

6.6.5 Design tools

Many tools are aimed at guiding the design process by allowing evaluation to inform design decisions. Such tools, by their nature, are often flexible in terms of the inputs and parameters that may be changed. This means that care must be taken to ensure that outputs are comparable when choices are being made between differing options. This is done through careful analysis of the boundaries and assumptions being made and the quality of the data being used. This may prove difficult when comparing results from a variety of sources. Many of the more advanced tools, such as dynamic energy-modelling packages, allow a high level of user input. This allows flexibility in developing different scenarios but can hide significant discrepancies in the quality of the underlying data and assumptions used.

Such tools will often include compliance outputs to allow them to be used to demonstrate compliance with regulations, BREEAM and other standards, but these specific outputs are based on a fixed, predetermined set of parameters based on the assumptions used in the regulatory tools, to ensure comparability.

6.6.6 Post-occupancy tools and processes

Use can be made of post-construction and post-occupancy tools and processes to ease the transition of a building into use, although this is in its infancy. Such systems can be very helpful in overcoming the disparity between design-stage predictions of performance and actual in-use performance. Occupant behaviour is a major factor in building performance, and through the provision of improved occupier support, informed by improved building-performance monitoring, very significant benefits can be achieved, in terms of both cost and wellbeing. Schemes such as Soft Landings in the UK (BSRIA, 2014a, 2014b), described in more detail in Chapter 7, provide a structured and externally monitored way of achieving these benefits, thereby giving confidence to all parties. The underlying principles can also be applied in a less formal way, although the benefits are often not as great.

6.7. Certification

A number of certification methods are available to provide an instantly recognisable evaluation of performance in the form of a score or rating. These have differing levels of rigour, and hence give varying levels of confidence in the final output.

■ *Third-party certification schemes* provide the highest level of confidence, as they are used by competent individuals. They are used to provide external verification of claims made about the performance of a product. Examples of such schemes are whole-building evaluation and certification methods such as BREEAM and LEED. Independently verified product certification schemes such as Category 3 environmental product declarations (EPDs) also fall within this category, although there is a need to ensure common boundaries apply in this instance, as the verification is simply made against the chosen criteria rather than any universal set of parameters.

■ *Self-certification tools* provide much less rigour, as there is no independent review. This leaves the potential for conflicts of interest, which can influence the reported outcomes, as well as the risk of blatant or unintended misuse unless there is a rigorous quality control process in place. Where it exists at all, the latter is normally a retrospective process, testing the competency of the person carrying out the process. Whilst the cost of applying self-certification tools is usually lower as they do not involve a third party, they are not suitable for external reporting or for auditing performance against requirements.

6.8. The tools

There are many tools available to assist engineers, design-team members, clients and others to evaluate performance against a wide range of environmental and economic impacts. There are fewer tools available to quantify social impacts, partially because these issues are by their nature less quantifiable and more subjective. However, the tools listed in the table in the Appendix do include certain social issues such as health and well-being and stakeholder engagement issues. The range of social impacts that can be measured is likely to increase as the tools develop over time. There are many single-issue tools available covering a wide range of individual design aspects such as energy perform-ance, daylighting and materials selection.

As already mentioned above, a number of established environmental standards for the built environment have been developed around the world, including the UK BREEAM, the US LEED Rating System, the German DGNB Certification System, the Australian Green Star Rating System, the Swedish Miljöbyggnad (MB) system, Pearl (Estidama) and the French HQE system. These are regularly updated, and details can be found on their respective websites.

Given the global influence of the US economy, LEED is seen by many as a pre-eminent global system due to its dominance of the large North American market, its growing global marketing reach and the influence of US-aid supported codification activity inter-nationally. However, LEED is arguably not as comprehensive or scientifically based as BREEAM, and is a less relevant standard in many situations due to a bias towards build-ing-services solutions, demands and standards commonly used in the USA. This can make it less challenging when applied to other climatic conditions. Internationally, LEED is increasingly perceived as less adaptable to local markets and regulatory systems.

Figure 6.2 shows an analysis of the relative number of categories and weightings of a number of established energy and environmental standards, as published in the book

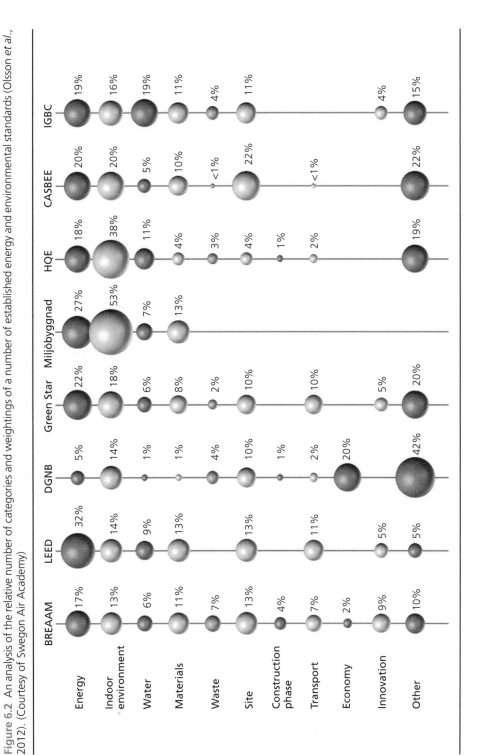

Figure 6.2 An analysis of the relative number of categories and weightings of a number of established energy and environmental standards (Olsson et al., 2012). (Courtesy of Swegon Air Academy)

Simply GREEN – A Quick Guide to Environmental and Energy Certification Systems for Sustainable Buildings published by Swegon Air Academy (Olsson *et al.*, 2012). (Note: The figure is based on the versions of the methods current at the time when the research was undertaken, and some of the methods have been updated subsequently. However, the underlying principles behind each method have not changed significantly between that research and the publication of this book.)

The Appendix at the end of this book provides further information on some of the key methods used in the UK and elsewhere, and maps these against the sustainability principles set out in the companion volume in this series (Ainger and Fenner, 2014).

6.9. Reporting performance

It is important that building performance is reported wherever possible in order to improve future practice. This is often done within individual companies, but in such a diverse sector as building construction it would be beneficial if more shared data were made available. In the UK, BRE, other organisations such as WRAP and the Carbon Trust and a range of cross-industry and government working groups are trying to fill this information gap but progress is slow due to the lack of available data on actual performance and costs.

Where performance is being used to compare projects externally, or to demonstrate achievement against contractual or statutory requirements, it is important that claims are properly substantiated using a credible third-party certification scheme. This is critical where information is being used within environmental or corporate social responsibility reports, or to bid for work, to ensure that the data are robust and comparable.

When reporting is used for other reasons, such as to guide internal decisions, management and continual improvement targets, an appropriate level of robustness needs to be determined. Where possible it is always preferable to use an independent assessment in such instances.

What can engineers do?

- Engineers have a major part to play in ensuring robust evaluation of buildings and the infrastructure around and supporting them. The principles outlined in this chapter provide a framework that will help to achieve this.
- Engineers can assist in the increased sharing of experiences and, most importantly, the public sharing and reporting of performance data, costs and other benefits. This is critical for improving the understanding of how buildings perform in reality, and how they can be improved. The use of industry standard tools and methods is central to this, as it provides the vital ingredient of comparability.

■ Engineers can use established tools to safeguard the interests of all parties. If carried out as an integral part of the design and construction process, this should not impose any significant burdens in terms of time or cost. Carrying out an evaluation too late in the process, especially where it is done retrospectively in order to justify earlier decisions, is likely to result in a poor outcome, unwelcome findings and significant expense in rectifying problems under severe time constraints. It is also likely to result in a less sustainable, or even an unsustainable, building.

REFERENCES

Ainger CM and Fenner RA (2014) *Sustainable Infrastructure: Principles into Practice*. ICE Publishing, London, UK.

BRE Global (2014) BES 6001 – The framework standard for responsible sourcing, Issue 3. BRE Global, London, UK.

BSI (British Standards Institution) (2007) BS 10621:2007 + A2:2012. Thief-resistant dual-mode lock assembly. BSI, London, UK.

BSI (2009) BS EN ISO 8902:2009. Responsible sourcing sector certification schemes for construction products. Specification. BSI, London, UK.

BSI (2010) BS EN 15643:2010. Part 1: Sustainability of construction works. Sustainability of buildings. General Framework. BSI, London, UK.

BSI (2011) BS EN 15643:2011. Part 2: Assessment of buildings. Framework for the assessment of environmental performance. Environmental impact assessment. BSI, London, UK.

BSI (2011) BS EN 15978:2011. Sustainability of construction works. Assessment of environmental performance of buildings. Calculation method. BSI, London, UK.

BSI (2011) PAS 2050:2011 Specification for the assessment of the life cycle greenhouse gas emissions of goods and services. BSI, London, UK.

BSI (2012) BS EN 15643:2012. Part 3: Assessment of buildings. Framework for the assessment of social performance. BSI, London, UK.

BSI (2012) BS EN 15643:2012. Part 4: Assessment of buildings. Framework for the assessment of economic performance. BSI, London, UK.

BSI (2012) BS EN 15804:2012. Sustainability of construction works. Environmental product declarations. Core rules for the product category of construction products. BSI, London, UK.

BSRIA (2014a) Soft Landings. Available at https://www.bsria.co.uk/services/design/soft-landings (accessed 12/11/2014).

BSRIA (2014b) Various documents on the website, including: *Soft Landings Core Principles* and *Soft Landings Framework*. Available at https://www.bsria.co.uk/services/design/soft-landings/free-guidance (accessed 12/11/2014).

CIRIA (2014) http://www.ciria.org (accessed 12/11/2014).

Commission for a Sustainable London (2012) *London 2012 – From Vision to Reality*. Post Games Report. Commission for a Sustainable London, London, UK.

European Commission (2002) Directive 2002/91/EC. Energy Performance of Buildings Directive. European Commission, Brussels, Belgium.

Greenhouse Gas Protocol (2013) *Protocol Life Cycle Accounting and Reporting Standard*. Available at http://www.ghgprotocol.org/files/ghgp/public/Product-Life-Cycle-Accounting-Reporting-Standard_041613.pdf (accessed 12/11/2014).

ISO (International Organisation for Standardisation) (2006a) ISO 14040:2006. Environmental management – Life cycle assessment – Principles and framework. ISO, Geneva, Switzerland.

ISO (2006b) ISO 14044:2006. Environmental management – Life cycle assessment – Requirements and guidelines. ISO, Geneva, Switzerland.

Millar M (1999) NASA says human error caused loss of Mars craft. *Yahoo! News*, 30 September.

Olsson D, Heincke C and Nilsson C (2012) *Simply GREEN – A Quick Guide to Environmental and Energy Certification Systems for Sustainable Buildings*. Swegon Air Academy, Kvänum, Sweden.

Part III

Change

Sustainable Infrastructure: Sustainable Buildings
ISBN 978-0-7277-5806-4

ICE Publishing: All rights reserved
http://dx.doi.org/10.1680/sisb.58064.113

Chapter 7
BIM: the sustainability context

Richard Shennan

The secret is to gang up on the problem, rather than each other.

Thomas Stallkamp

7.1.   Introduction

Creating or improving buildings is a complex business involving many skills and disciplines, usually housed in a diverse range of businesses right through the supply chain. The impact of the decisions made during those processes have a profound impact on the performance of the building with respect to its meeting its social objectives, its environmental impact, and its operating cost and underlying value over the life cycle – in other words the sustainability of the building. BIM is a new way of working, enabled and driven by technology, but founded on improved process and more effective collaboration. Well-organised and shareable information is at the heart of BIM, with all parties able to use a common data set for the purpose that they require. It allows the barriers that exist between diverse stakeholder groups to be removed. It has the potential to bring the same degree of improved efficiency to the design, construction and performance of the built environment as has been achieved already in manufacturing and process industries. These efficiencies and other benefits of BIM can make a major contribution to the design, construction and operation of buildings that address the three pillars of sustainability.

BIM enables interaction and collaboration between designers, constructors, owners, users and facilities managers to deliver the desired outcomes. The principles of effective collaborative working are set out in the British Standard Code of Practice for Collaboration, BS 1192:2007, and subsequent PAS 1192-2:2013 and PAS 1192-3:2014 published by the British Standards Institution (BSI, 2007, 2013, 2014).

7.2.   The importance of information exchange

Much of the information that is generated by the participants over the project cycle and the knowledge which that can foster is lost in the two major information-exchange stages: that from the designer to the constructor, and that from the constructor to the owner. The organisation and representation of the information, and the technology used to exchange and finally deliver it, has not moved on substantially in 30 years. It is little surprise, therefore, that when they are set to work in reality so many buildings fail to achieve the performance targets imagined by the early-stage concept designers.

BIM, and the technology that supports it, can enable the step change in performance of the built environment that is so badly needed for a sustainable future. BIM in itself will not ensure a sustainable outcome but is a key enabler that can significantly enhance the likelihood of achieving this.

7.3. Delivering better outcomes

Every building project has an intended outcome. Sustainable buildings can only be achieved if the outcomes are properly defined at the project-definition stage and understood by all parties. Outcomes should be defined in terms of the following aspects:

■ the social reason for the building
■ the financial benefits and value over the life of the building
■ the environmental impact.

Once these have been put in place, BIM becomes a process that connects the intended outcomes all the way through the design and construction stages to the actual performance of the building over its life. By continuing to manage the data and the information that it provides, a feedback loop can also be created that helps to create learning and continuous improvement of outcomes.

7.4. Improving existing buildings

If we are to meet the wider challenge of sustainability, substantial improvement in the performance of existing buildings is also required. This can now be achieved by capturing spatial information and available data for existing buildings, and moving it into a BIM environment. Laser-scanning technology is readily available and affordable, creating accurate point cloud data that can be transferred into the BIM software environment. Available data can then be added to create a common data set, and standard software can be applied to allow the benefits of the BIM process to be applied to existing buildings in the same way as for new buildings. In addition, wasteful repeated surveys of the same building for various purposes over its life can be eliminated.

7.5. Federated models

The most widely used configuration of an information model is known as a 'federated model' (Figure 7.1). This comprises a set of interconnected models that are fully coordinated and make use of a common data environment, where data are held and referenced in a way that allows all parties to be able to access information in a structured way. In its fullest sense, an information model includes graphical and non-graphical data, and related documentation, configured and interlinked, with clear lines of provenance and change control.

7.6. Enhanced early-stage optioneering

Early-stage option appraisal is the essential first stage in sustainable building design, both for new buildings and improvements. The predicted building performance as determined against the desired outcomes needs to be assessed for various options at an early stage, before architectural design solutions begin to narrow down. This requires the use of specific predictive software in areas such as efficient structural solutions, computerised

Figure 7.1 A federated model. (Image courtesy of Mott MacDonald)

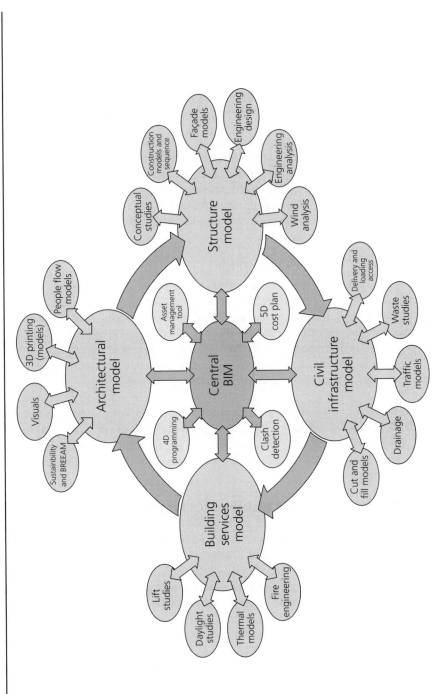

fluid dynamics, through-life cost and carbon modelling, comfort, water consumption and compliance with particular functional parts of a brief. If these predictive design packages are used in isolation, there is a labour-intensive requirement to enter the base data for a given option, so that the number of options that a designer can afford to consider within a given fee budget is limited. Using isolated models also leads to the substantial risk that the data input for one analysis can become out of sequence with design developments in other 'silos'.

A federated model structure can allow effective data transfer from spatial to analytical models, and a well-managed common data environment can ensure that the various studies are coordinated. This means that a lot less time is wasted on data input and obsolete analysis, which in turn means that more options can be considered in a short time and optimisation can be carried out across a series of weighted parameters. For a given time and cost, the final solution will be closer to the absolute optimum against the defined sustainable outcomes.

In practical terms this does mean that all parties need to construct their early-stage models with appropriate levels of detail to allow the information exchange to take place effectively. Too much detail in an early-stage model can clog up the flow, and is not consistent with the iterative nature of an optimised design process. Careful planning of the process and the increasing levels of development of models is essential. Outline stage models need to be developed at an appropriate level of dimensional tolerance, avoiding the temptation to start adding detail and to look at dimensional accuracy too early simply 'because you can' when using typical BIM-enabling software.

Software developers are increasingly working towards this objective, and a key element of a project BIM execution plan is to look at all potential purposes of the model, based on the employer's information requirements, identify the required level of development for each application, and then select software platforms that allow this early-stage infor-mation exchange.

7.7.   Visualisation and stakeholder engagement

Delivery of a sustainable building requires a close match to be achieved between the objectives of the owners, users and operators, and the delivered project. Early-stage brief-ings and sign-offs are commonly used in attempt to achieve this, but the media of two-dimensional plans or relatively superficial architectural renderings can often be a barrier, especially to those stakeholders who are not used to interpreting drawings and schedules as a depiction of the design intent. Moving into a model-based environment can bring the presentation of the proposals to life, although designers must be careful to clarify that high-quality visualisations at outline stages do not imply that all design issues have been resolved. By clearly communicating design intent at an early stage, feedback can be obtained and incorporated in the next level of development of the model. Physical 3D 'prints' (e.g. that of Victoria Station, London, in Figure 7.2) can also be used to commu-nicate design intent. As well as helping with the wider definition of sustainable outcomes, use of visualisation linked to the BIM process can address some of the more practical issues, such as ease of maintenance and simple control logic.

Figure 7.2 A 3D model image of the Victoria Station upgrade. (Courtesy of Mott MacDonald, Taylor Woodrow/Bam Nuttall JV and London Underground)

Early-stage visualisation can also be used to engage wider stakeholders, such as local communities who may be affected by a proposed building development, politicians and funders.

7.8. Building performance modelling

Early-stage building performance modelling brings great benefits to the appraisal of options. BIM processes can take this a step further, especially in connection with managing change. Under current methodology, a lot of investment is made in building predictive performance models at the outline design stage, but when activities such as 'value engineering' take place, decisions are often made that undermine key aspects of the original strategy. This often happens once contractors have been appointed, or when budget problems arise. The originally intended performance starts to become detached from the project as constructed.

By developing a model-based design, and maintaining that model as contractors and the supply chain get involved, it is much easier to extract and exchange the information required to re-run the predictive performance models to make sure that sustainability is not compromised.

7.9. Embodied carbon data as a BIM 'dimension'

A BIM model can house a wide range of attributes associated with the objects that make up the model. These attributes are grouped into classes, generally referred to as 'dimensions' of BIM. So the three spatial dimensions is the obvious starting point (3D), the fourth dimension is time (4D) and the fifth dimension is cost (5D). The full potential is often referred to as nD. One such dimension is embodied carbon. By linking the material specifications and quantities that are inherent in a BIM model to a database of embodied carbon, a rapid assessment can be made of the total embodied carbon of a particular solution. This then brings carbon intensity into play as a parameter that can be applied to the early-stage option appraisals described above, and can also allow a visual representation of carbon intensity to be developed. For those building procurers that wish to drive down overall carbon emissions associated with their project, a visual presentation, for example using simple colour-coding to show intensity in the various objects that make up the design model, can allow them to understand where they may be able to review other requirements that they may have set as design parameters, in order to include the carbon dimension in their decision-making.

7.10. Reduced waste

Current design, construction, operation and maintenance methodologies generate significant waste in many respects. Time and effort is wasted during the design stages due to out-of-sequence working and inconsistency in the information that is being used by different members of the design team to carry out their particular tasks. Although the value of time does vary widely around the world, ultimately wasted endeavour leads to lost opportunity to save more tangible physical resources. Waste of a more material kind is generated as a result of coordination problems that are only uncovered during the construction itself, leading to physical rework and its associated waste. A lack of accuracy in construction information leads to over-ordering, which again raises the likelihood of waste.

The overall scale of waste can be even greater during the operation of the building over its life cycle. By keeping an up-to-date set of asset information and designing for ease of maintenance using a 3D model, objects within assets will have an extended useful life. Complete replacement is often a decision that is taken in the context of a lack of access to the information required for effective maintenance.

7.11. Design for manufacture and assembly

It has long been recognised in the construction industry that off-site fabrication can bring major benefits in terms of quality, cost and waste. These parameters are also metrics for sustainable buildings. There are also associated benefits in terms of health and safety. BIM is a catalyst for a rapid increase in the use of design for manufacture and assembly (DfMA), with building sites becoming partly places of assembly rather than just raw construction.

Building information models that become a virtual prototype of the construction process, able to look into spatial, time and cost dimensions, make it much easier to identify and analyse opportunities to use DfMA. The particular elements identified can have multiple components, including structure, building services and finishes, where appropriate, due to the use of a single federated model that integrates the different disciplines.

The application of DfMA can lead to a substantial reduction in the total materials used and in carbon emissions, both embodied and in transporting materials to and from site.

A single, integrated component can be delivered to site, instead of numerous separate component deliveries and subsequent removal of surplus materials.

The model can be used accurately to plan points of connection on site, and clashes can be checked and avoided. The relevant parts of the integrated whole-building model can be extracted and passed to the fabricator, who can then add additional detail as required and pass their DfMA component back for checking within the primary model. This method of working also requires a collaborative and trusting relationship between project stakeholders.

By using the power of 4D BIM, the state of a partially constructed building can be clearly visualised based on the construction programme and proposed delivery schedules for the off-site components, and the on-site assembly process can be planned and optimised in the model.

By introducing cost information into the model, it can be significantly quicker to look at the potential cost impact of selecting composite elements for DfMA, which helps to test the case.

The benefits of component-based design, with multiple suppliers providing integrated elements extracted from a central model, which are then assembled to form a complete product, are well recognised in other industries. However, the construction industry has been quick to point out that this approach will not work due to the unique nature of each building. The emergence of BIM provides the opportunity for use of models to plan much more accurately, and actually to see at an early stage in the design process where the benefits of DfMA can be obtained, and thereby to address risks related to ill-planned, ill-fitting prefabricated components, which have impeded uptake of DfMA to date.

7.12. Soft Landings

Soft Landings is an approach first developed by BSRIA and the Usable Buildings Trust based on an original idea by Mark Way and now adopted by the UK Government Construction Board as Government Soft Landings (GSL). It focuses on the optimum performance of buildings in use, and outcomes in terms of the building's purpose. The use of BIM to provide a single data set that is continuous through the project life makes it a key facilitator for successful implementation of Soft Landings, allowing facilities managers to collaborate more effectively during the design and construction stages. Soft Landings should, therefore, be an overlay to BIM-based process maps, with the required activities taking place at each stage of development of the model, from concept through to completion and continuous commissioning.

The power of BIM to assist visualisation can be deployed to convey design intent to those who will be responsible for operation and maintenance of the building, at a stage in the process when their views can still make a difference, thereby helping to fulfil a key aspect of Soft Landings.

7.13. Integrated asset information management

Asset management systems vary in complexity according to the scale of a building and the number of building assets under common ownership or management. Common to all such systems is a requirement to have access to the information needed for operation and maintenance, and a need to measure performance and interpret any divergence from the optimum. This, in turn, leads to actions needed for sustainable outcomes, in terms of the ability of the building to deliver its social function, the life-cycle cost and the environmental impact.

BIM gathers and organises data, and thus generates accurate information. The data are key for the owners and operators, as some of the detailed requirements of the design and construction process become less relevant. In order to achieve integrated asset information management over the life cycle of a building, there are four key stages:

- At the outset of the BIM process, consider the data and information that owners and operators will actually need, and plan the model development so that this is incorporated in the common data environment.
- As the design and construction stages progress, build the data set and plan the exchange of information with the owner in a way that is compatible with its future use.
- At the point of handover, exchange information in an agreed format. The UK Government's preferred information-exchange platform at present is COBie (Construction Operations Building Information Exchange), which was originally developed in the USA and is based on a relatively simple Excel format, can import data from design- and construction-stage models, and has the potential for the data to be transferred into facilities management systems.
- As the asset is put into use, and perhaps modified or refurbished over its life cycle, the data set should be maintained and updated with information gleaned in the field to allow continuous commissioning towards optimum performance.

Asset management software is now embracing the potential of BIM to bring a new dimension to the information based on current technology. The 3D spatial models developed as part of the BIM process are being used as the point of access to the information. Virtual representations of the building can be called up on portable devices such as tablet computers, and used as a gateway to the information required for effective operation and maintenance. Increasing use of radiofrequency identification (RFID) tags or data-matrix barcodes on equipment could allow the site operative to get confirmation of the details of components so that they are sure of what they are doing. Physical operation and maintenance manuals will soon be replaced by accurate and up-to-date information, accessed through a visual model, giving operatives the correct information, when they need it and where they need it. Systems also increasingly provide a facility to submit updates from the field, to enable the central asset model to be kept up to date.

BIM offers great potential to get better performance from buildings in use, which is a key objective for sustainable buildings.

7.14. Performance optimisation

BIM has the potential to allow a much more informed operation of a facility, with the target performance set against all parameters that can be understood and used as a target for sustainable operation and best value.

If clear performance targets are set out they can be incorporated in the common data environment and viewed alongside the data recorded in the field. The way in which the data are represented, and the information that they convey, can be set up to match the actual operation and maintenance. This enables a layer of engineering interpretation to be applied, looking at divergence, exceptions and waste, and then feeding the information back into model so that field operatives can see the requirements and take effective action to achieve optimised performance against all selected parameters.

7.15. Feedback and learning

The lack of connection between the information generated at the design and construction stages and the information generated and used during operation and maintenance has wider consequences than the inefficient performance of a single building asset. The ongoing development of high-performance buildings, whether new or refurbished assets, requires a continuous process of feedback and learning. The history of the development of sustainable buildings is far from linear. New ideas are often developed by designers and can be affected by trends that gain substantial traction in the design community, but there can be a lack feedback and participation from owners and operators. There are many examples, ranging from overcomplicated control systems to technologies that simply do not deliver the savings in reality. Successful innovations can be slow to propagate and differentiate themselves from less effective initiatives. An example of success is the Elizabeth Fry Building at the University of East Anglia (Figure 7.3), which was designed in 1991, using simple diurnal heat storage in the thermal mass of the building, combined with a high-performance envelope. This building is still one of a handful in the UK that has maintained its target level of energy consumption over 20 years later.

There are two primary reasons for the lack of a continuous process of feedback and learning: the lack of consistent information collected from buildings in use and its assessment against the intended performance; and the tendency for the involvement of designers to come to an end when assets are handed over, so that they do not learn from the performance of their own designs. This has recently been recognised in the new *RIBA Plan of Work 2013* (RIBA, 2013) with the introduction of Stage 7 – In Use. In the middle are the contractors, who typically work in a procurement system that leads them to focus on the handover of the asset at the completion of their work, and move on.

BIM can address the first of these problems, by providing a clear and organised set of data that can enable the post-occupancy assessments to be carried out in a much more effective way. The second problem may require a more fundamental shift in the procurement system and the roles of construction professionals, with a greater emphasis on risk and reward sharing relating to performance in use, and potentially a greater involvement of those involved in design in providing an ongoing advisory role, helping owners, operators and facilities managers to interpret and synthesise information within a

Figure 7.3 The Elizabeth Fry Building – the best building ever? (Courtesy of the *Building Services Journal*)

common data environment in a way that brings benefits to the owners as well as reducing the environmental impact of buildings in use.

7.16.   Decommissioning and recycling

The final stage of the life cycle can also benefit from BIM. The existence of an accurate and accessible set of data for a building can lead to a better planned decommissioning. Design models can begin to incorporate a dimension that is focused on recycling of materials, in the same way that other industries already have. For existing buildings, hazardous materials can be identified and recorded in a model-based environment, which will give ready access to the relevant information in a visual format. These potential benefits are at an early stage, but once the industry adopts BIM as the default means of designing, constructing and operating buildings, it will be able to follow and become a standard part of the new process.

7.17.   Summary

The UK construction industry is preparing for a step change in the application of technology to deliver benefits that are long overdue. The adoption of BIM by the UK

Government (BIM Task Group, 2014) has raised the profile of the methodology and put the industry on alert that change is required. Sustainability is one of the key drivers of this policy. But the potential benefits are now being recognised widely in the private sector as well, thus accelerating the change. The catalysts for this change are twofold: the new economic reality, which requires increasing social demands on the built environment to be met with less consumption, less cost, and less carbon; and the maturity of technology that is bringing the potential for efficiencies already seen in other industries into construction.

Buildings are not mass produced, and each building tends to have unique factors, but the current state of technology means that this can no longer be used as an excuse. The primary issue likely to dictate the rate at which BIM can be deployed in support of a sustainable built environment is the readiness of the industry to embrace change, develop new collaborative working relationships, engage the supply chain and the facilities managers in the process, and establish outcomes in use as the primary focus of the efforts of all that work in the industry. It is also essential that building owners and procurers recognise and fulfil their roles in clearly identifying their requirements and preparing themselves to make best use of digital asset information over the life cycle of the building.

What can engineers do?

- BIM is basically about well-organised asset information that is shared through a common data environment.
- Collaborative working between all project partners is at the heart of BIM.
- The greatest potential value of BIM is over the life of the building.
- Information exchange at the appropriate level of development should be captured on a process map as part of project planning.
- Technology is an enabler, not a definer, of BIM.
- BIM can be applied to any project, and there is no reason not to manage information exchange and collaborative working for common benefit on any size of project or with any organisation.

REFERENCES

BIM Task Group (2014) Available at http://www.bimtaskgroup.org (accessed 12/11/2014).
BSI (British Standards Institution) (2007) BS 1192:2007. Collaborative production of rchitectural, engineering and construction information. Code of practice. BSI, London, UK.
BSI (2013) PAS 1192-2:2013 Specification for information management for the capital/ delivery phase of construction projects using building information modelling. BSI, London, UK.
BSI (2014) PAS 1192-3:2014. Specification for information management for the operational phase of assets using building information modelling. BSI, London, UK.
RIBA (2013) *RIBA Plan of Work 2013*. RIBA Publishing London, UK.

Sustainable Infrastructure: Sustainable Buildings
ISBN 978-0-7277-5806-4

ICE Publishing: All rights reserved
http://dx.doi.org/10.1680/sisb.58064.125

Chapter 8
Future direction of measurement methodologies

Alan Yates

Progress is the activity of today and the assurance of tomorrow.
Ralph Waldo Emerson (American Poet, lecturer and essayist, 1803–1882)

8.1.   Introduction

The role of, and principles behind, the robust evaluation of sustainability through the procurement process and beyond was discussed in Chapter 6. While many of the tools and policies in use today attempt to promote forward-looking actions, they are necessarily limited by our lack of knowledge of what the future holds and the necessity to work within current practices, technologies, knowledge and economic limitations. As a result, much of our current 'forward-looking' design is, in reality, often driven by policies and targets with a short time horizon of no more than 5–10 years. Given the life expectancy of our buildings, a time horizon of well over 50 years would be more appropriate if we could determine suitable targets with a reasonable degree of confidence.

These tools are there to service and support design, construction and operational processes, not to replace them. It is vital that they facilitate rather than dictate solutions, and so promote good practice and robust innovation from all involved in building procurement and operation. The need to base this on sound science and evidence is paramount, as is the need to balance the wide range of often conflicting issues discussed in Chapter 6.

Buildings are among the longest lived of the assets in our built environment. While a few may be designed to last for a short period of time, most will be designed to last for at least 30 years, and in reality usually last for much longer. Given the resources that are involved when constructing a building, we should be designing buildings to last well beyond their notional design life and current functional use, as well as ensuring that the resources embedded within them can be easily and efficiently reclaimed for future reuse wherever possible and recycled where not. In the UK, HM Treasury uses a life expectancy of 50 years in its *Green Book* (HM Treasury, 2011), which governs public-sector procurement, including buildings and infrastructure. The BRE uses a notional building life of 60 years in its life-cycle assessment work for typical building types based on long-standing research into building life, while certain components, such as concrete structures, are at least theoretically capable of lasting considerably longer than this if other factors do not result in them becoming unfit for purpose and so leading to demolition. In reality, while we would like to think that buildings currently under construction will last, history shows that lifespan is unpredictable, as it depends on a diverse range of factors well outside our control and often unconnected with the physical durability of the building or its materials.

Regardless of this dilemma, assessment and evaluation methods need to keep abreast of the likely changes in use of buildings over their lifetime, attempt to forecast the future needs and technologies where possible, and encourage flexibility and adaptability where it is not. This chapter explores some of the key issues that will impact on the way we design, construct and, therefore, measure the success of our buildings in the future.

8.2.   Drivers for change in the building sector

On average, UK citizens spend approximately 90% of their time in buildings and much of the remaining time travelling between them. Thus the built environment and the connecting buildings and infrastructure are highly influential on economic value, our society, its health and its vibrancy, as well as the wellbeing and satisfaction of the individuals who use the buildings.

The importance of buildings makes them a key component in planning for change. For this reason there is a need to ensure that they retain or increase their value to us over their life cycle. As changes happen in society, commerce, health and leisure, the technologies that support buildings will continue to evolve, and our buildings will need to adapt accordingly. Property owners, pension and investment funds and occupiers are increasingly aware of the role that buildings play in these areas, and it is vital that resilience of the building stock is maintained.

There are a number of key areas that will affect our buildings in the future, and these need to be taken into account in the development of design processes, evaluation tools and methods. These key areas are discussed below.

8.2.1   Resilience and adaptability – energy consumption and its impact on climate change

The subjects most often discussed as impacting on future building design policy are global warming and the resultant risks of climate change. While the science is not fully understood and the scale of change is far from clear, most scientists and policy makers agree that it is appropriate to take precautionary actions to reduce harmful emissions of carbon dioxide and other potent greenhouse gases, and to take climate change into account when designing, constructing or refurbishing buildings.

The biggest risks to the built environment come from temperature and precipitation changes resulting from changes in the atmosphere and the global air and marine currents that govern our climate (in the UK, warmer, wetter winters and, potentially, warmer, drier summers), increased flooding, greater storm frequency and intensity, and an increased possibility of building instability through changes to local hydrological patterns.

Buildings are a prime factor contributing to the reduction of greenhouse gas emissions. In the UK, they are currently responsible for approximately 50% of the total national emissions of carbon dioxide. This is primarily through heating and cooling, lighting and other electrical power used within them. Policies to reduce the dependency of our built environment, and society more generally, on fossil fuels will impact on the former, but much of this energy use relies on electrical generation, which is outside the scope of the building itself. While renewable energy sources will have a significant part to play, and their use where appropriate is to be encouraged, this alone will not achieve the targets being set for reductions in operational carbon emissions.

Global, European and national policy decisions over recent years have increased the pressure to reduce energy consumption significantly, and in particular the associated carbon emissions. The UK Government set a target to achieve 'zero carbon' performance by 2016 for residential buildings and 2019 for non-domestic buildings in England. Their definition of 'zero carbon' is still being developed, and has changed significantly since the policy was first announced. It now only covers so-called 'regulated energy' (in simple terms, the energy used by fixed equipment for heating/cooling, permanent lighting, ventilation and some small power in some building types). The definition is likely to include the ability to offset a proportion of the carbon emissions through 'allowable solutions', although these have yet to be fully defined and a system put in place to measure and audit them. However, the target is still challenging, and for some does not seem financially viable. Certainly zero carbon buildings will not be achieved (even accepting the current government definition) without a major advance in industry practices and technologies, which will entail rapid change in the run-up to these target dates. 'Business-as-usual' approaches will not get us there.

In reality, many buildings at present under construction fall well behind current regulation in terms of their performance. This is largely due to the relatively long lead-in times involved in their delivery, and for this reason the stated target dates are likely to be somewhat more relaxed. Nevertheless, there is no room for complacency in adopting low-energy design and construction principles, and the engineering and architectural professions have a key role to play in implementing them.

Decarbonisation of the National Grid is a key component in achieving the targets but this falls well outside the influence of the building sector and its many stakeholders, although engineers have a key role to play in this as well. An important issue to consider is how far a building should go to reduce its reliance on grid electricity if the grid itself will be decarbonised in the foreseeable future. While there is no simple answer to this, decisions should be based on the premise of minimising demands through built form and construction, maximising efficiencies in the system, and then taking actions to meet such demands in a low-carbon manner where possible.

8.2.2   The design versus performance gap

Design-stage calculations, whether for regulatory purposes, certification or to guide design decisions, are universally based on prediction tools. These are based on assumptions relating to use patterns and occupant behaviour taken from historical data. This raises questions over their use to predict actual in-use energy bills. Nonetheless, they are valuable as comparative tools to allow design decisions to be made, tested and audited, as long as they provide a clear direction and scale to the performance that can be expected from a building.

Having said that, it is clearly important that prediction tools reflect reality insofar as this is practically possible. This means improving the way that we predict user behaviour, and in particular how this might change in the future. Greater use of control systems can help to achieve better regulation of system performance and economy but their popularity and robustness varies considerably from one building type to another. Evidence indicates that many control systems installed in buildings are either not used in practice, due to a lack of understanding of their operation with those managing the property, or are actively bypassed to achieve 'desired' results, so negating the potential benefit. Coupled with the need for regular maintenance and calibration – which often does not happen as intended – this means that such strategies are often not as successful as they could be.

The use of more complex dynamic energy and thermal modelling software helps the designer, as it allows a design to be tested rapidly against a range of climate and user scenarios. Today this is widely done for more complex buildings.

Improved software coupled with greater use of the systems' capabilities will undoubtedly help to improve evaluation tools used in regulation and certification in the future. However, many buildings rely on much simpler methods, and financial limitations will not permit the use of complex modelling on such projects in the near future. This means that the simpler models will need to be developed further as our understanding of user behaviour improves.

The increasing use of BIM, as outlined in Chapter 7, should allow more robust modelling techniques to be adopted over time but improved feedback on actual building performance is needed to validate the calculations and assumptions behind them. A variety of UK organisations, including the Carbon Trust, CarbonBuzz, BRE and others, are attempting to collate data, and it is important that engineers play their part in collecting and sharing these data.

Research is currently underway by the UK Government, led through the Zero Carbon Hub and other non-governmental organisations, as well as by a range of national research bodies, including the BRE and a number of universities, to identify the potential for improving the regulatory and design models used. This is likely to impact significantly on industry practice over the next few years.

8.2.3   Risks arising from higher building performance

The drive to increase building performance is inevitably creating the risk of conflicts and unforeseen consequences occurring. Higher energy performance inevitably requires greater levels of airtightness in a building. While this will be less of a problem for a fully air-conditioned building, it is likely to cause health and comfort difficulties and materials/fabric deterioration in other building types unless adequate controlled ventilation is provided and steps are taken to avoid overheating. This, in turn, raises concerns over the ability of building users and managers to understand and operate the solutions provided to overcome these risks. Often these are bypassed, either due to ignorance, to save the perceived energy costs associated with their operation, or due to the apparent lack of individual control over the internal environment.

Many design practices and design tools deal with issues individually and fail to make these connections. They will need to develop and become more integrated to overcome this situation if problems are to be avoided.

8.2.4   The growth of community infrastructure solutions

As the need for higher performance levels and lower costs increases, greater consideration will be needed of the potential benefits of using community or neighbourhood solutions for heating/cooling, energy, waste, surface water management and water collection, green space and amenity. This will inevitably bring building and local infrastructure and urban design closer together. Design practices and assessment tools will need to be adapted to be able to recognise the benefits of such an approach. BREEAM is expanding its scope to address and recognise these issues, and is working with the Civil Engineering Environmental Quality (CEEQUAL) certification scheme to coordinate the approach being taken on local and larger scale infrastructure projects.

8.3.   Changes that will influence building evaluation in the future

Generally speaking, technology and processes evolve in response to the need to meet a changing demand, and the built-environment sector is no exception. While new technologies are often expensive at the outset, these costs tend to decrease rapidly as the technologies become more mainstream.

It is not possible to list all of the potential technological and process changes here, but they fall broadly into the following categories:

8.3.1   Building fabric solutions

■ The greater use of modular systems and off-site construction methods to improve resource efficiency, reliability and timeliness of construction, to enhance quality and to reduce costs.
■ Improved manufacturing of materials and components to reduce the embedded resource, social, ethical and economic impacts of the products themselves and to reduce their life-cycle impacts.
■ Use of reactive materials and components to allow the building envelope to minimise solar and climate impacts on the internal environment. The increased use of nanotechnologies in materials manufacture over the coming years will lead to high-performance materials capable of achieving significantly higher levels of thermal performance and solar control than is currently possible, in a cost-effective way.
■ Improved technologies and techniques to allow for the use of thermal mass to temper and regulate internal environments through the appropriate use of high mass/thermal storage where a stable environment is required, and lower thermal mass where a more responsive environment is appropriate as a result of intermittent occupation.

- Greater use of passive design solutions to minimise demands, whether this is through a fully passive design or in a mixed-mode context, where passive and active systems are adopted to achieve an optimum outcome.
- Increasingly more flexible design and construction to allow for enhanced adaptation to climate change and building function during a building's life.
- A greater emphasis on designing for deconstruction to allow greater reuse and recycling of materials, thereby reducing the impact of the building across its overall life cycle.

8.3.2   Improved building systems

- Development and use of higher efficiency systems and components.
- Renewable energy technologies and other low carbon solutions such as biomass, combined heat and power (CHP) or district heating/cooling networks, etc.
- Improved control and monitoring systems, including the greater use of demand-reduction technologies aimed at shifting energy use to periods of lower demand, thus reducing peak energy demands on the National Grid.
- Use of smart technologies to allow external monitoring and control of building systems by both energy suppliers and premises management companies.
- Designing systems to allow for efficient and appropriate upgrading of performance to meet changing demands as the building use evolves over time. This does not mean a simple overspecification of a system, such as the provision of larger capacity boilers/chillers/fans to allow for increased demand should it ever happen. Such overspecification is unlikely ever to be properly utilised, and will result in inefficient operation throughout the life of the system and unnecessary use of resources through the provision of a larger system than is necessary.

8.3.3   External changes

- Changes to national energy infrastructure, including the decarbonisation of the National Grid, the use of carbon capture and storage systems and the use of local community-level energy networks.
- Enhancements in local transport infrastructure to reduce dependency on high-impact modes of transport such as the use of fossil-fuel-powered private cars, through improved and lower-impact public transport, enhanced cycling provision and the improved safety of pedestrians and cyclists.
- Demographic changes such as rapid population growth and increasing affluence are especially important in many parts of the world and will have a dramatic impact on the way in which the built environment develops. The major developing economies (including the so called 'BRIC' and 'MIST' countries of Brazil, Russia, India, China, Mexico, Indonesia, South Korea and Turkey) are growing rapidly and present an opportunity to achieve high performance and low impact from the start. Unfortunately, they also present a significant risk that these opportunities will be missed in the pursuit of rapid and easily achieved results using typical current 'developed world' practices. Given the role of consultants and suppliers from the established 'old world' economies in this development, there is a part for them to play in ensuring that this balance is skewed towards low impact solutions.
- Increased globalisation of manufacturing has an important influence on the ability of individual nations to set and work towards their own goals. It is vital that this

does not result in the growth of a 'one size fits all' mentality, given the varying local needs and opportunities presented by climate.

■ Other social and economic factors arising from demographic changes, business needs, societal demands and expectations, including changing health needs, household sizes, workplace requirements, employment and economic growth.

■ Improved understanding and quantification of social and socio-economic impacts arising from the built environment.

■ Greater emphasis on the role that the built environment, and in particular our homes and workplaces, play in improving health and wellbeing, thus reducing the burden on society's health services, which are traditionally reactive rather than preventive.

■ Improved understanding of the impacts that buildings have on the environment, society and economy, including the impacts of occupant behaviour and management practices. Enhanced feedback mechanisms and greater post-handover support and monitoring are key to achieving this.

All the factors listed in Sections 8.3.1 to 8.3.3 are likely to influence the criteria, methods and metrics that we use to evaluate building performance in the future. For this reason, evaluation tools will continue to evolve and adapt to take account of growing knowledge.

8.3.4 International standards relating to whole-building performance

Over recent years work has been undertaken by CEN to develop European framework standards for the sustainability assessment of buildings (see Chapter 6). This work is still ongoing and, while voluntary, these standards will impact on the future of whole-building evaluation methods in Europe (Box 8.1).

The biggest changes will occur in relation to the environmental analysis and declaration of impacts through environmental product declarations (EPDs) and materials specification tools such as the BRE *Green Guide to Specification* (BRE Global, 2014) and Impact (Impact, 2014). Greater harmonisation of these tools will greatly improve the ability of designers and specifiers to compare options and so make informed decisions.

Box 8.1 CEN European framework standards

■ BS EN 15643. Sustainability of construction works. Sustainability of buildings.
Part 1:2010. General framework.
Part 2:2011. Assessment of buildings. Framework for the assessment of environmental performance. Environmental impact assessment.
Part 3:2012. Assessment of buildings. Framework for the assessment of social performance.
Part 4:2012. Assessment of buildings. Framework for the assessment of economic performance.
■ BS EN 15804:2012. Sustainability of construction works. Environmental product declarations. Core rules for the product category of construction products.
■ BS EN 15978:2011. Sustainability of construction works. Assessment of environmental performance of buildings. Calculation method.

While these standards will impact on the future of whole-building evaluation tools such as BREEAM and DGNB, these changes are not likely to be that significant, as the standards are based on a need to harmonise rather than improve, and both of these methods reflect the broad aims of the standards. Other national schemes may be more dramatically impacted by these framework standards if they choose to comply. LEED and other international schemes will be less likely to comply as they are not focused on the European market. The standards are most likely to impact on the way in which the outputs of such tools are presented, and lead to an increased use of standardised metrics, which is to be welcomed.

8.4.   Changes in building procurement processes

The way in which we design and procure buildings in the UK will continue to change significantly over the next few years, and this is likely to be mirrored in other countries, if not globally. A change in contractual arrangements and the effect on the roles and responsibilities of those involved in the process can have a major impact on the quality and performance of the final product. A move away from the traditional adversarial contractual relationships to a more collaborative approach would be a major step forward in working towards the challenges that we face. It would allow greater sharing of information and objectives, and encourage increased use of shared information to ensure a joined up and fully integrated approach to building design.

Such changes, coupled with the greater complexity of our buildings and their systems, and the recognition that improved building management is required, are driving towards the use of more integrated design processes and the use of BIM systems to improve coordination and information transfer. While these changes are reactive, they will have a dramatic change on the way in which decisions are made, tools are developed and operate (Box 8.2), and information is stored and evaluated and results are presented. The full range of evaluation tools will need to adapt to take account of these changes.

> **Box 8.2** Changes in building procurement processes
>
> BRE is carrying out research on the impact of such changes on BREEAM and the *Green Guide to Specification* (BRE Global, 2014), and Impact (Impact, 2014) is designed specifically with BIM in mind. Such research by the BRE and others will undoubtedly change the way in which regulation and certification systems work in the future, and should help to reduce significantly the burdens that these systems can create for design teams, constructors and their clients, while improving the outcomes that they achieve. This is probably the biggest single medium-term change that will occur in building evaluation. While technological evolution can be easily accommodated within its existing structure, BREEAM and its fellow tools and methods will evolve, and are unlikely to look or feel the same as they develop to take account of these process changes.

8.5.   The value of evaluation

Evaluation of building performance is, and will continue to be, vital to achieving quality, efficiency and reliability. As building solutions become more complex and external challenges more onerous, the need for such tools to aid decision-making, inform detailed specification, guide construction, and inform and guide building operation will increase.

A key component to improving these tools is the better understanding of the impact of their use, and the benefits and value that they bring. A focus on better feedback mechanisms is vital to this. Value, both in the short term and in longer life-cycle financial terms, is central to this. Recent research by the Royal Institute of Chartered Surveyors (RICS) and Maastricht University (RICS, 2012) shows the impact that building certification has on property values, with some surprising results. The research showed property value increases of up to 20% for highly-rated buildings, and equivalent research on LEED-assessed buildings in the USA gave similar results (Kok, 2011). The BRE has carried out research into the costs of BREEAM ratings in both capital and whole-life terms (Abdul *et al.*, 2014). While this research is in its infancy it is seen as critical to informing the ongoing development of their assessment methods.

8.6.   Conclusion

The future is unpredictable. However, the longevity of buildings and the value of the resources and effort embedded within them mean that we must not be complacent about the need to design and construct for change.

While technology can meet some of the changing demands, ultimately it is down to our ability to design, construct and operate buildings in a manner that facilitates change over time through adaptability and flexibility. The processes that we use at these life-cycle stages to manage design, construction, maintenance and refurbishment also have a critical part to play in achieving the desired outcomes. These will change rapidly over the next few years as we improve our ability to integrate design solutions and share accurate information on buildings and their performance. Increasingly, buildings will operate as a part of a broader system in terms of energy data and the social, economic and environmental environment more generally, bringing in broader issues of urban and utility infrastructure, which will need to be carefully considered.

Building evaluation tools will need to evolve significantly over the coming years as technologies and processes change. While this might well change the appearance and accessibility of such tools and the ways in which they are used, the need for such tools will increase in order to facilitate robust and credible specification and evaluation of performance.

While greater harmonisation is likely to occur in some areas, such methods and tools are driven by the market and its needs. The need for methods and tools varies both between sectors and globally, and as such a 'one size fits all' approach is not an option. Neither should this result in an overstandardisation of the design solutions adopted, which would detract from the overall quality and interest of the built environment.

Local standards based on recognised international frameworks provide the best way forward. They provide a level playing field, enhanced understanding with improved skills and a clear direction of development for an increasingly global industry.

It is only by ensuring that outcomes are monitored and that feedback mechanisms are in place within the design and construction sectors that the providers and developers of evaluation tools can ensure that their products remain fit for purpose and responsive to the needs of the industry.

What can engineers do?

The future development of tools guidance and data is dependent on the participation of all stakeholders in the building procurement and management sectors. Centrally-collected, globally-relevant, accessible benchmarking data are necessary to give relevance to the assessment, whichever method is used. Through the ongoing collection and sharing of data on real projects, and feedback to the developers and operators of these media, engineers can have a significant impact on the future direction of this work.

Professional bodies will have a role to play in the future direction, and members can use their involvement to influence this. Most importantly, engineers can ensure that issues are properly considered in their work at the most appropriate time, and that decisions are made with due consideration of the life-cycle impacts, rather than being made purely on the basis of short-term objectives.

Perhaps most importantly and positively, engineers can play their individual part in innovating and pushing boundaries to tackle the rapidly changing demands on the buildings and their associated infrastructure.

REFERENCES

Abdul Y, Quartermaine R and Sutton D (2014) *Delivering Sustainable Buildings – Savings and Payback*. IHS BRE Press, London, UK.

BRE Global (2014) *Green Guide to Specification*. Available at http://www.bre.co.uk/greenguide/podpage.jsp?id = 2126 (accessed 12/11/2014).

BSI (2010) BS EN 15643:2010. Part 1: Sustainability of construction works. Sustainability of buildings. General Framework. BSI, London, UK.

BSI (2011) BS EN 15643:2011. Part 2: Assessment of buildings. Framework for the assessment of environmental performance. Environmental impact assessment. BSI, London, UK.

BSI (2011) BS EN 15978:2011. Sustainability of construction works. Assessment of environmental performance of buildings. Calculation method. BSI, London, UK.

BSI (2012) BS EN 15643:2012. Part 3: Assessment of buildings. Framework for the assessment of social performance. BSI, London, UK.

BSI (2012) BS EN 15643:2012. Part 4: Assessment of buildings. Framework for the assessment of economic performance. BSI, London, UK.

BSI (2012) BS EN 15804:2012. Sustainability of construction works. Environmental product declarations. Core rules for the product category of construction products. BSI, London, UK.

HM Treasury (2011) *The Green Book – Appraisal and Evaluation in Central Government*. HM Treasury, London, UK.

Impact (2014) Integrated Material Profile and Costing Tool. Available at http://www.impactwba.com (accessed 12/11/2014).

Kok N (2011) Supply, demand and the value of green buildings. Paper presented at the 47th Annual AREUEA Conference, Chicago, IL, USA.

RICS (Royal Institute of Chartered Surveyors) (2012) *Supply Demand and the Value of Green Buildings*. RICS, London, UK.

Sustainable Infrastructure: Sustainable Buildings
ISBN 978-0-7277-5806-4

http://dx.doi.org/10.1680/sisb.58064.135

Chapter 9
Envoi

Tristram Hope

> It was strange there, in the very depths of the town, with ten miles of man's
> handiwork on every side of us, to feel the iron grip of Nature, and to be
> conscious that to the huge elemental forces all London was no more than the
> molehills that dot the fields.
>
> Sir Arthur Conan Doyle

This second volume in the *Delivering Sustainable Infrastructure* series has investigated a wide variety of issues ranging from the simple to the highly complex. In many respects, the writing of the book and the absorption of the ideas by the reader are the easy bits. The real challenge is to take the thinking that has been described and implement it on everyday projects, in the face of scepticism and in spite of the ever-present commercial pressures that drive the construction industry in all corners of the globe.

We have no choice but to engage with the process of changing our direction away from rampant consumerism and towards a more moderate, responsible way of doing things. We have one world in which to live, and, if we continue to abuse it, that world will not come to an end, it will merely carry on regardless – without us.

It is our choice as human beings how marginal we allow our existence to become, and we have no one else to blame but ourselves for our own inability to act for the greater good. We can decline into a 'dog eat dog' attitude, or we can battle to rise above it. Furthermore, as Sir Nicholas Stern pointed out in his 2006 *Review of the Economics of Climate Change*, the longer we prevaricate, the harder and more costly the battle will be, so the time to start is now.

In producing this book, the drafting team were driven by the sincere hope that the information contained within it will be of interest, will provide encouragement and, above all, will be of practical use in assisting the reader in the fickle process of designing better and more sustainable buildings in the future.

We have given you the tools: it is now for you to get out there and begin the job. Good luck!

REFERENCE

Stern N (2006) *Review of the Economics of Climate Change*. HM Treasury, London, UK.

Part IV

Tools

Sustainable Infrastructure: Sustainable Buildings
ISBN 978-0-7277-5806-4

ICE Publishing: All rights reserved
http://dx.doi.org/10.1680/sisb.58064.139

Appendix 1
Practical resources

Alan Yates

> The expectations of life depend upon diligence; the mechanic that would perfect his work must first sharpen his tools.
>
> Confucius

Chapter 6 sets out the underlying principles that need to be considered when selecting appropriate tools to guide the building designer. This appendix provides a list of key methods currently in general use and maps them against the sustainability principles set out in the companion volume in this series (Ainger and Fenner, 2014).

The list is, however, far from being complete, as there are many tools available from a wide variety of sources, including many specialist and commercially focused tools. Many of these may be well developed and credible, although others may not be. Some may be biased towards specific design or specification solutions.

In addition to those listed below, there are many resources available that relate to specific sectors, product lines or solutions, such as:

- carbon footprinting
- ecological footprinting
- renewable energy generation potential
- flooding
- daylight and lighting levels
- energy consumption
- life-cycle costing.

It is not possible to list these here, given the number available and the rate of change that occurs. The tools included in the table below are well established and have a history of successful application, a high level of peer review and a reasonable degree of independence. They are largely, but not entirely, UK based.

Whatever tools and guidance you use, it is important to bear in mind that they are only of value when they are appropriate to the task in hand and the context in which they are being used. As a general rule, the results will only be as good as the user's knowledge and skill. The output is sensitive to the quality and accuracy of the input. It is, therefore, important to ensure both that the tools you are using are appropriate and that you are using them correctly. This is especially true where tools are sophisticated in nature. The ill-informed use of such tools presents a high risk of inaccuracy, and potentially increased liabilities, because the risks and consequences of misuse may be significant and often only manifest themselves much later downstream when rectification becomes difficult, expensive and disruptive. Significant reputational and commercial damage can also result.

It is, therefore, always advisable to consult practitioners with experience before using a tool for the first time or in a new context. This will ensure that you understand its limitations and how best to make use of it to guide your decision-making.

REFERENCE

Ainger C and Fenner RA (2014) *Sustainable Infrastructure: Principles into Practice*. ICE publishing, London, UK.

Scheme/tool	Details/source for further information	Absolute principles			
		A1. Environmental sustainability – within limits	A2. Socio-economic sustainability – 'development'	A3. Intergenerational stewardship	A4. Complex systems
Whole-building evaluation tools					
BREEAM (Building Research Establishment Environmental Assessment Method) (UK, Netherlands, Spain, Norway, Sweden, Germany and international versions) ■ New construction: – non-domestic – domestic ■ Refurbishment and fit-out: – non-domestic – domestic ■ BREEAM In Use (existing buildings) ■ Communities ■ Infrastructure (under development)	BREEAM was first launched in 1990 and has evolved in both scope and coverage. It is now used across a number of European countries through nationally adapted versions. It is also applied through a common international version for use in all countries where a national version does not exist. BREEAM adopts the relevant principles and scope of the European framework standards relating to the sustainability assessment of buildings published by CEN (see Box 6.6 in Chapter 6, and Box 8.1 in Chapter 8). BRE (Building Research establishment, UK): http://www.breeam.com DGBC (Dutch Green Building Council): http://www.breeam.nl ITG (Fundacion Instituto Technológico de Galicia, Spain): http://www.breeam.es DIFNI (Deutsches Privates Institut für Nachhaltige Immobilienwirtschaft, Germany): http://www.difni.de NGBC (Norwegian Green Building Council): http://www.ngbc.no SGBC (Sweden Green Building Council): http://www.sgbc.se	✓	✓	✓	✓

Scheme/tool	Details/source for further information	Absolute principles				
		A1. Environmental sustainability – within limits	A2. Socio-economic sustainability – 'development'	A3. Intergenerational stewardship	A4. Complex systems
Code for Sustainable Homes (New-build homes only) (England, Wales, Northern Ireland only)	The Code for Sustainable Homes is a derivative of BRE's EcoHomes method developed by the BRE and the UK Government's Department for Communities and Local Government (DCLG). It is owned by the UK Government, and is currently under review with a view to being wound down. The BRE is developing a UK domestic scheme within the BREEAM family of tools to promote more sustainable new homes into the future. http://www.planningportal.gov.uk/buildingregulations/ greenerbuildings/sustainablehomes	✓	✓	✓	
LEED (Leadership in Energy and Environmental Design) ■ Building design and construction ■ Interior design and construction ■ Building operations and maintenance ■ Neighbourhood development ■ Homes	LEED was derived from BREEAM and launched in 2000. Like BREEAM it has evolved in response to changes and needs. The methods make use of US codes and standards, although some flexibility is allowed to adapt to local needs when used internationally. US Green Building Council; http://www.usgbc.org	✓	✓	✓	✓

	Description					
DGNB (Deutsche Gesellschaft für Nachhaltiges Bauen/German Sustainable Building Council) (Germany, Denmark)	The DGNB system adopts the relevant principles and scope of the European framework standards relating to the sustainability assessment of buildings published by CEN (see Box 6.6 in Chapter 6, and Box 8.1 in Chapter 8). DGNB: http://www.dgnb-system.de	✓	✓	✓	✓	✓
Green Star (Australia, New Zealand, South Africa) ■ Design and as built ■ Interiors ■ Existing buildings (performance) ■ Communities	Green Star was derived from BREEAM and provides similar coverage of issues. Green Building Council Australia: http://www.gbca.org.au/green-star New Zealand Green Building Council: http://www.nzgbc.org.nz Green Building Council South Africa: http://www.gbcsa.org.za	✓	✓	✓	✓	✓
Esitdama – Pearl Rating System (Abu Dhabi) ■ Buildings (office, retail, multi-residential, school, mixed use) ■ Communities ■ Villa	The Pearl Rating System is mandated in Abu Dhabi for many building projects and provides similar coverage to BREEAM and LEED. It is based on the four pillars of the Estidama vision, and provides ratings at the design, construction and operational stages in a building's life cycle. http://www.estidama.upc.gov.ae	✓	✓	✓	✓	✓
Passivhaus (Germany, UK, various others)	Passivhaus promotes a low-energy 'fabric first' approach to building design based on a prescriptive set of building performance parameters. http://www.passivhaus.org.uk	Partial	Partial	Partial	Partial	Partial
NABERS (National Australian Built Environment Rating System)	NABERS is a national rating system that measures the energy efficiency, water usage, waste management and indoor environment quality of Australian buildings, tenancies and homes. http://www.nabers.gov.au/public/WebPages/Home.aspx	Partial	Partial	Partial	Partial	Partial

Scheme/tool	Details/source for further information	Absolute principles			
		A1. Environmental sustainability – within limits	A2. Socio-economic sustainability – 'development'	A3. Intergenerational stewardship	A4. Complex systems
CASBEE (Comprehensive Assessment System for Built Environment Efficiency) (Japan)	Like BREEAM, LEED, etc., CASBEE provides an in-depth evaluation of a building's performance at the pre-design, design and post-design stages. It is developed specifically for use in the pre-design, new construction, existing building and refurbishment life-cycle stages. Japan Sustainable Building Consortium: http://www.ibec.or.jp/CASBEE/english/overviewE.htm	✓	✓	✓	✓
Energy tools					
SAP (Standard Assessment Procedure) (UK)	SAP is the UK national calculation method for homes, and is used in the Building Regulations for England, Scotland, Wales and Northern Ireland. http://www.gov.uk/standard-assessment-procedure	Partial		Partial	
SBEM (Simplified Building Energy Model) (UK)	The UK national calculation method for Non-domestic buildings – Used within Building Regulations in England, Scotland, Wales and Northern Ireland http://www.ncm.bre.co.uk	Partial		Partial	

Materials tools

Green Guide to Specification (UK)	The *Green Guide* is an accessible life-cycle assessment based guide to help designers and specifiers make early choices about the specification of materials for key building elements such as external walls, floors, roofs windows and cladding systems. BREEAM uses the *Green Guide* and Impact to award credits under its materials section. BRE: http://www.bre.co.uk/page.jsp?id=499	Partial	Partial
Impact (UK)	The Impact methodology is an integrated materials profiling and whole-life costing tool. BREEAM uses the *Green Guide* and Impact to award credits under its materials section. Impact: http://www.impactwba.com BRE: http://www.bre.co.uk	Partial	Partial	Partial
BEES (USA)	Building life-cycle assessment software developed by the National Institute of Standards and Technology (USA). http://www.nist.gov/el/economics/BEESSoftware.cfm	Partial	Partial	Partial
Athena Impact Estimator (USA/North America)	A whole-building tool covering a range of environmental impacts arising from materials specification. It produces a cradle-to-grave inventory for a building and allows the regional context to be taken into account. The tool covers new construction, refurbishments and extensions in the North American market. Athena Sustainable Materials Institute: http://www.athenasmi.org/our-software-data/impact-estimator	Partial	Partial	Partial

Scheme/tool	Details/source for further information	Absolute principles			
		A1. Environmental sustainability – within limits	A2. Socio-economic sustainability – 'development'	A3. Intergenerational stewardship	A4. Complex systems
Athena EcoCalculator (USA/North America) ■ Commercial buildings ■ Residential buildings	A structured Excel-based tool with a built-in library of common constructions and materials impact data. It is structured for various North American markets. Athena Sustainable Materials Institute: http://www.athenasmi.org/our-software-data/ecocalculator/	Partial	Partial	Partial	
Indoor air quality tools					
Healthy Development Measurement Tool (USA)	San Francisco Department of Public Care	Partial		Partial	
Design quality tools					
Building for Life 12 (UK)	The Building for Life scheme provides a set of underlying principles to be considered during design and an independent peer-review process. It is often referred to by planning authorities and some developers in the UK. The scheme is not comparable to any of the whole-building evaluation methods listed above as it does not provide a quantification of impacts or a means of setting targets, benchmarking or reporting performance. The outcomes are based on subjective peer review. The Design Council: http://www.designcouncil.org.uk/knowledge-resources/guide/building-life-12	Partial	Partial	Partial	

Waste			
DoWT-B (Designing out Waste Tool for Buildings (Waste and Resources Action Programme (WRAP), UK)	A tool to be used in conjunction with WRAP's guidance: Designing out Waste – a Design Team guide for buildings. It helps in identifying opportunities, comparing and recording solutions, calculating impacts and providing indicative waste forecasts. http://www.wrap.org.uk/content/designing-out-waste-tool-buildings	Partial	Partial
SMARTWaste	SMARTWaste is a construction site waste-management tool with a number of variants to help in the evaluation and benchmarking of construction waste impacts. It is developed by BRE in collaboration with major UK construction companies. http://www.smartwaste.co.uk	Partial	Partial

Principles for sustainable infrastructure

Engineers understand the concept of physical principles when solving an engineering problem. Similarly, in order to test the sustainability of an action, it is important to set choices and decisions against guiding principles of sustainability. These play a key role in setting the context for the choices that organisations make, and whilst some focus on values, others prompt a defined method or standard for implementation. The first book of this series, *Sustainable Infrastructure: Principles into Practice*, develops and expands the principles in Table A1 and discusses their application at each planning and implementation stage of an infrastructure project.

Table A1 Principles for sustainable infrastructure. (From *Sustainable Infrastructure: Principles into Practice* (Ainger and Fenner, 2014))

Objectives			Goals, approaches

Absolute principles
These are incontrovertible and are consequences of natural science laws and basic humanity.
*They represent **constraints** within which infrastructure services must be delivered.*

A1	A2	A3	A4
Environmental sustainability – within limits	Socio-economic sustainability – 'development'	Intergenerational stewardship	Complex systems [Edwards' list of principles]

Operational principles
These are more specific to help set objectives and guide day-to-day practice.
They help to establish a distinct way of doing things, in part by recognising issues that may not have been part of traditional remits.
*They guide the **processes** that can be adopted at appropriate stages of each project.*

O1.1	O2.1	O3.1	O4.1
Set targets and measure against environmental limits	Set targets and measure for socioeconomic goals	Plan long term	Open up the problem space
			O4.2 Deal with uncertainty
O1.2 Structure business and projects sustainably	O2.2 Respect people and human rights	O3.2 Consider all life-cycle stages	O4.3 Consider integrated needs
			O4.4 Integrate working roles and disciplines

Individual principles
It is individuals, acting alone or collectively, that make decisions and influence a more sustainable project outcome
*They reflect **values**-based aspects of sustainability.*
They cut across all four absolute principles, reflecting ethical and professional responsibility as well as individual personal behaviour.

I1 Learn new skills – competences for sustainable infrastructure

I2 Challenge orthodoxy and encourage change

Sustainable Infrastructure: Sustainable Buildings
ISBN 978-0-7277-5806-4

ICE Publishing: All rights reserved
http://dx.doi.org/10.1680/sisb.58064.149

Index